THE TRANSCENDENT SCIENCE

NIJHOFF INTERNATIONAL PHILOSOPHY SERIES

VOLUME 15

General Editor: JAN T.J. SRZEDNICKI
Assistant Editor: LYNNE M. BROUGHTON

Editorial Advisory Board:

For a list of other volumes in this series see final page of the volume.

Clark Zumbach

The Transcendent Science

Kant's Conception of Biological Methodology

1984 **MARTINUS NIJHOFF PUBLISHERS**
a member of the KLUWER ACADEMIC PUBLISHERS GROUP
THE HAGUE / BOSTON / LANCASTER

Distributors

for the United States and Canada: Kluwer Boston, Inc., 190 Old Derby Street, Hingham, MA 02043, USA
for all other countries: Kluwer Academic Publishers Group, Distribution Center, P.O.Box 322, 3300 AH Dordrecht, The Netherlands

Library of Congress Cataloging in Publication Data

Zumbach, Clark.
  The transcendent science.

  (Nijhoff international philosophy series ; v. 15)
  Includes bibliogrpahical references and index.
  1. Biology--Philosophy. 2. Kant, Immanuel, 1724-1804
--Science. I. Title. II. Series.
QH331.Z86 1984   574'.01   83-23678
ISBN-13: 978-94-009-6106-7   e-ISBN-13: 978-94-009-6104-3
DOI: 10.1007/ 978-94-009-6104-3

ISBN-13: 978-94-009-6106-7

Copyright

For Ezekiel and Andrew

Contents

Preface

The most neglected sector of Kant's Critical Philosophy is his collection of remarks about biological phenomena in the second part of the *Critique of Judgment*, the Critique of Teleological Judgment. The reasons for this are numerous, but since in Kant, everything comes in threes, a three-fold collection will suffice. The Critique of Teleological Judgment itself is one reason. More than most of his writings, this segment of the Critical corpus suffers from what can most charitably be termed "mistakes of exposition." In this part of the third *Critique*, it is commonplace to find sub-arguments in Kant's general position somewhere other than their logical niche. The result is that the general theme behind his remarks about living phenomena is obscured. This difficulty has done much to discourage even the most enthusiastic of Kant admirers from investing their time on this work. Secondly, in this century, until very recently, there has been little interest in philosophical questions about biology. Twenty-one out of thirty-one sections of the Critique of Teleological Judgment (sections #61 and 63-83) deal either directly or indirectly with issues of interest in the philosophy of biology. Finally, the Critique of Teleological Judgment has been placed among the last on that list of writings thought to formulate Kant's Critical system. This is not merely because of its temporal position. (The *Critique of Judgment*, published in 1790, was chronologically later than the first two *Critiques*, the *Prolegomena*, the *Groundwork of Metaphysic of Morals*, and the *Metaphysical Foundations of Natural Science*.) It has also been frequently assumed, I think mistakenly, that its logical place within the Critical system is among the last. The result is that very few readers of Kant have spent much time on the Critique of Teleological Judgment; there is very little knowledge of it *per se*, let alone of how it fits into the scheme of the Critical Philosophy.

The present study has in part been motivated by this neglect of the Critique of Teleological Judgment. For it is essentially a work on a certain thread running through the Critical Philosophy: Kant's philosophical remarks about biological phenomena. Since most of Kant's remarks on this topic occur within the Critique of Teleological Judgment, most of my attention will be directed towards this part of the *Critique of Judgment*. More specifically, I will be concerned with sections #61 and 63-82 of the Critique of Teleological Judgment. The Critique of Teleological Judgment, as I have come to realize, is really a Critical examination of teleological judgments in general. But I shall, as I have already indicated, be focusing on only one of these: the judgment that something is living.

In a study such as this it is customary to play down at the outset the influence other contributors to the secondary literature have exerted in the formation of the author's finished project. To a large extent I do not have to worry about this tendency since there has been so little written on the Critique of Teleological Judgment. Still the Critique of Teleological Judgment is a difficult piece, and to understand it, I have not hesitated to make use of an acknowledge the work of others, be it by writers on Kant or writers in philosophical biology. But when points of difference arose, I also tried, as much as possible, to mention these as well.

While writing this book I have been fortunate to receive both scholarly and philosophical help from many. I am most greateful to my friend, Peter Kivy. Special appreciation must also be expressed for the help generously given to me by Ralf Meerbote, W.H. Walsh, and John Yolton. I am also grateful to Allen Ginsberg, Alan Hart, Robert Matthews, Richard B. Miller, Mark Siderits, Patric Stayton, W.D. Stine, Mark Timmons, Robert Van Gulick, and Robert Weingard for studying and evaluating earlier stages of parts of this work. Finally, I wish to thank Debra for her patience and encouragement.

Reference to Kant's works

All references to the *Critique of Pure Reason* are given by citing either or both pages of the two original editions, A and B. With occasional minor alterations, the Kemp Smith translation is used.

All references to Kant's *Critique of Judgment* will be to (1) the section and page of this work in Vol. V of the *Akademie* edition, abbreviated "Ak," with the first number indicating the section and the second the page (The *Akademie* pagination appears in the margin of the Meredith translation of this work.), and (2) the corresponding page in J.H. Bernard's translation of the *Critique of Judgment* (New York: Hafner, 1974), abbreviated "B." All quotations are taken, with frequent modification, from the Bernard translation.

References to Kant's other translated works will follow the same format as that used in citing the *Critique of Judgment*. These works are:

> *Critique of Practical Reason* (Vol. V of the *Akademie* edition), trans. L.W. Beck (Indianapolis: Bobbs-Merrill, 1956), abbreviated "BkR."
>
> *First Introduction to the Critique of Judgment* (Vol. XX of the *Akademie* edition), trans. James Haden (Indianapolis: Bobbs-Merrill, 1965), abbreviated "H."
>
> *Groundwork of the Metaphysic of Morals* (Vol. IV of the *Akademie* edition), trans. H.J. Paton (New York: Harper and Row, 1964), abbreviated "P."
>
> *Prolegomena* (Vol. IV of the *Akademie* edition), trans. L.W. Beck (New York: The Liberal Arts Press, 1951), abbreviated "Bk."

The Doctrine of Virtue: Part II of the Metaphysic of Morals
(Vol. VI of the *Akademie* edition), trans. Mary J. Gregor
(Philadelphia: University of Pennsylvania Press, 1964), abbre-
viated "G."

I will occasionally modify these latter translations.

CHAPTER I

Teleological phenomena

that the cause or causes of order in the universe probably bear some remote analogy to human intelligence...

David Hume

1. Teleology and reduction: Preliminaries

By virtue of its very title and my preparatory remarks, it is perhaps expected that the Critique of Teleological Judgment contains an extensive discussion of the teleological character of living things. Indeed, it is generally thought that the Critique of Teleological Judgment has something to do with teleology and biology. Although this is to some extent true, I shall argue that Kant's remarks about the teleological character of living things have an import which has been almost totally overlooked. In the present section I want to lay bare the skeleton of my approach to Kant's interesting work. However, inasmuch as the teleological character of living things will figure prominently in what follows, a preliminary exploration of this feature of the world would be helpful at this point.

In more recent discussions of teleology, it has been customary to distinguish two sorts of features of living things as teleological. These are the functional and the goal-directed characteristics of living things. For what follows, it will be useful to have this distinction before us, even if it is drawn in only a loose and intuitive way.

When a function is ascribed to a part in a living organism, one is indicating that the part has an effect of a certain *sort*. Red blood cells have the *function* of carrying oxygen throughout the body; carrying oxgen throughout the body is an *effect* of red blood cells.

The question which those giving an analysis of function statements attempt to answer is that of how we are to distinguish effects of something which are functional from those which are not. For a human being in an extremely cold climate, having a nose through which one can breathe is extremely important. Were one to breathe through one's mouth under these conditions, frostbite of the lungs would result. The nose's effect of acting as an intake for air in the human organism allows for the air to warm-up somewhat before it enters the lungs. This sort of effect of the nose, being an intake for air in the human organism, is a function of the nose. However, one of the nose's effects is its capability of supporting eye glasses. Yet most philosophers and biologists would disagree with Voltaire's Pangloss and deny that this is a function of the nose. But again, whether this effect is to be counted as a function will depend on what criteria for functionality we adopt. To complicate matters further, biologists, as philosophers seldom realize, distinguish between functions which are necessary and those which are not. And by this their intent is not simply to distinguish those *effects* of something which are necessary from those which are not. Thus, one might say with a gleam in his eye that acting as an intake for air to the human body is a necessary function of the nose. Serving as a support for eye glasses is a function of the nose which is not necessary. Finally, being a site where the adolescent frequently gets pimples, is an effect of the nose which is completely non-functional. What makes functions "teleological" is that there seems to be some sense to citing these sorts of effects of parts in biological systems to account for the part's presence. Whether or not there is in fact sense to such explanatory patterns is a question which goes hand in hand with the question of what sort of effect is to count as functional.

The concept of function begins to get fuzzy when one tries to distinguish from it a second teleological concept, that of goal-directedness. Whereas an activity is functional in virtue of *being a certain sort of effect of a part* in a living system, an activity is goal-directed in virtue of *the manner in which it is controlled*. For example, a human's normal body temperature is 98.6° F. If the body gets over-heated or chilled, then sweating or shivering occur in order to bring the body back to a normal temperature. The body's "goal" is to maintain this normal state in spite of the various changes in its internal and external conditions. This maintenance of normal temperature is said to be an activity which is goal-directed. There are, of course, functional activities in this process: for example, sweating is a function of sweat glands. Nevertheless, the maintenance of a normal temperature is not something which is said to be functional. Such activity is described, on the face of it, in such a way that it sounds as if the goal of the activity, the normal state, is guiding or controlling the activity. This is how goal-directed processes become clouded with a teleological aura. But biological mechanisms actually control goal-directed processes in living things. A living process which is goal-directed is one which is directly under control of some other trait of the living system, be it a part of the brain controlling the body's temperature or a cell's DNA directing protein synthesis.

But just at the point where one is beginning to get an intuitive feel for the distinction, certain questions can no longer be restrained. Why cannot we think of the maintenance of a normal temperature as a function of the body? Why is not the nose's effect of serving as an air intake to the body an instance of goal-directed activity? More-over, strictly speaking, functional activities are also the result of a controlling mechanism; the structural proteins which compose the nose are effects of the DNA of various types of cells. Thus, some sort of directional mechanism is relevant to functional, as well as goal-directed activities.

Yet notice that the relevance of the controlling mechanisms of living organisms is not quite the same in both teleological activities, and perhaps herein lies that which distinguishes the two. In central cases of ascribing functions to parts of living systems, one need not make reference (either implicitly or explicitly) to the controlling or directing mechanisms which lead to the functional effects. If one wants to know the function of the mitochondria in a eukaryotic cell, one need only indicate that the function of a cell's mitochondria is to serve as centers for ATP generation, and leave it at that. The mechanisms which bring about the function need not be mentioned. To do this would be to enter into the etiology of a functional effect. However, to describe any activity as goal-directed is to thereby imply the presence of some controlling or directing factor not simply as part of the etiology of the activity, but as an essential part of the activity.

Although not a few philosophers will disagree with the way I distinguish functional and goal-directed activity, drawing the distinction in this way brings two aspects of living processes to the surface. In a living thing one can find those features which *control* other features and those features which *are controlled by* other features. We might say that there are those features of a living thing which are designer-like and those which are the result of these designer-like processes. For example, molecular genetics distinguishes one group of chemicals which carry the genetic information and control cellular processes, the nucleic acids, from a second group, the proteins, which are controlled by the nucleic acids.

Now Kant does not explicitly draw the function/goal distinction. But the distinction between that part of the organism which is the planner and that part which is the result of the plan does figure prominently in his remarks about the philosophical problems living phenomena present. In the remainder of this section I will show how.

It is possible to distinguish physical objects, as Kant does implicitly in the Critique of Teleological Judgment, into three types: man-made artifacts, inorganic objects, and living organisms. The division between the designer-like mechanism and the pattern occurring as a result of this mechanism is applicable to artifacts. Artifacts have a designer and their functional features are the result of design. But in the case of most artifacts, the designing mechanism is *external* to the artifact. The wooden fence built by my father is the result of a designing and directing mechanism. But no part of the fence itself designs or directs any processes which lead to the structural features of the fence. The sophisticated feedback mechanisms of the twentieth century of which Kant had no knowledge carry out a *portion* of the directing and controlling activities of their human creators. However, this division is not applicable to inorganic objects. Neither is there a part or aspect of a stone which directs or controls other features of the stone, nor are there any parts of the stone which are the result of any directing or controlling processes. Finally, there are living things. It is an observable fact that a living thing has, as Kant says, "*formative* power and to be sure such a formative power that is communicated to its materials which lack it (they are organized); therefore it has a self-propagating formative power..." (Ak: 65, 374; B: 221). A living thing has two aspects: the designer-like aspect and the designed-like aspect. Thus, it is possible to distinguish two features of living phenomena which can also be distinguished in the case of artifacts, but which are not distinguishable in inorganic objects. (Kant seemed to think that one exception to this distinction was ecosystems. They are clearly not artifacts, but they also neither seem to be mere inorganic objects nor living things, though they do

exhibit a designed-like aspect in that their components are functional. A virus also seems to be an exception to this three-fold division.
Viral particles do have a designer-like aspect; they contain their own
nucleic acid. But since they have no metabolic system of their own,
it might be most accurate to say that they have no designed-like
aspect. Yet we shall see that nothing depends on these possible
exceptions to this tripartite division.)

To Kant, this two-fold aspect of living things not present in inorganic matter suggests that both cannot be *viewed* as the same kind
of physical thing. Indeed, he does more than just suggest this. It is his
belief that while the behavior of inorganic phenomena can be "explained" by physical science, living processes, because they manifest
these two aspects not present in the latter, cannot. In his words: "It
is indeed quite certain that we cannot even become sufficiently
knowledgeable of, much less provide an explanation of organized
beings and their internal possibility according to mere mechanical
principles of nature..." (Ak: 75, 400; B: 248). The mechanical conception of nature, which as we shall see, is for Kant, just the physicist's framework of nature, leaves the behavior of biological phenomena in some significant sense "inexplicable."

This reveals the issue with which Kant's remarks about the teleological character of living things are directly concerned. They are not
made in response to the question of whether the universe and certain
things in it are the product of intelligent design.[1] He is not, that is,
arguing for a version of "physicotheology," which as he puts it, "is
the endeavor of reason to infer the supreme cause of nature and its
attributes from the *purposes* of nature (which can only be discerned
empirically)" (Ak: 85, 436; B: 286-7). Thus, although Kant believes
that the living things, or as he refers to them, "organized beings"
cannot be explained in terms of the "mechanical principles of
nature," this is not to be taken as the claim that there is intelligent
design in their etiology. Moreover, he is not (as is more often
thought) *directly* concerned with what philosophers have called the
"problem of teleology," the question of whether or not it is *legitimate* to explain the presence of something in a living system teleologically, i.e., in terms of its function(s) or the goal of the entire system.[2] Rather, Kant's discussions of teleological description and explanation have as their object the Cartesian claim that biology is just
a chapter in physics. Kant is here concerned with the suggestion
made by some seventeenth and eighteenth century thinkers that all

vital phenomena are explicable exclusively in terms of the physical sciences.[3] The central philosophical issue of the Critique of Teleological Judgment then is neither the claims of physicotheology nor the question of whether teleological talk in the science of living phenomena is legitimate. It is rather, as commentators are frequently concluding about other areas of the Critical Philosophy, the issue of reductionism.[4] In this case, it is the question of whether living phenomena are reducible to highly complex, highly special patterns of physical processes.

Admittedly, the question of the reduction of biology to physics is not unrelated to the problem of teleology. If one can show that all teleological talk in the biological sciences is legitimate from the physicochemical perspective by indicating that it is translatable into purely physicochemical descriptions, and that teleology is the distinguishing mark of living phenomena, then one has shown that there is a sense in which the "reduction" of biology to physics is possible. Conversely, if teleological judgments are not translatable into non-teleological ones, then the reduction of biology to physics seems to be thwarted. But to read Kant as being concerned exclusively with the question of the *legitimacy* of teleological judgments results in overlooking another issue to which this question is inextricably connected, namely, the question of reductionism. More specifically, to ignore that here the central issue for Kant is the reducibility of biology to physics, results in overlooking Kant's sensitivity to the various concepts of reduction that arise in our attempts to characterize biological methodology.

On the basis of what has already been said, one might infer that Kant will be taking a steadfast anti-reductionist stance on the question of whether biological processes are nothing over and above physical ones. However, in Chapter IV we shall see that Kant distinguishes three different concepts of reductionism: "ontological reductionism" (or "physicalism"), "methodological reductionism," and "explanatory reductionism."[5] It will in turn become clear that Kant argues against only the latter sort of reductionism; Kant is an *"explanatory anti-reductionist."* The extent and significance of Kant's refusal to subsume living phenomena under the kind of object studied by the physical sciences — a refusal which, as I have indicated, is motivated by the fact that biological processes lend themselves to the above teleological distinction — is the central object of study in this work.

2. Purposiveness and designedness

In addition to things which are the result of conscious design, there are entities in the universe which appear to be the result of conscious design. Human artifacts are things we know to be the result of conscious design; their existence is known to be causally determined in part by the conscious intent of human agents. Other entities in the universe, such as ecosystems and biological organisms, have an order which makes them appear as though they too are the result of conscious design. But in our more cautious moods, we say that ecosystems and biological organisms only *appear* in this way. We make this more guarded claim about these natural things because *prima facie* there are no grounds sufficient for establishing that they are in fact the result of conscious design. Granted both natural entities and human artifacts are susceptible of functional characterization, perhaps even in the same sense of the word "function;" nevertheless the fact that we can in principle verify experientially that human artifacts are products of intentions and cannot do this with respect to natural things permits us to say that at best the latter appear to be the result of conscious design.

Given that some natural entities appear to be the result of conscious design, the difference in our knowledge of the origin of such natural things can be marked conceptually. We might say that these natural entities indicate the presence of what Kant in the *Critique of Judgment* and elsewhere calls "purposiveness."[6] The purposiveness indicated by some natural things is to be contrasted with what might be termed the "designedness" confirmed by human artifacts.[7] We know that one of the causal conditions which determines the order found in artifacts is that of conscious intent. It is in virtue of this causal condition that an artifact establishes the presence of designedness. We cannot say then that either ecosystems or biological organisms point to designedness in nature, at least not until we can produce some sort of evidence that their origination involves conscious intent. Artifacts, on the other hand, *do not* indicate the presence of mere purposiveness, but designedness. ("Designedness" then, as I shall be interpreting Kant, does not entail "purposiveness.")

One might object to Kant's contrast between natural things which appear to be the result of conscious design and artifacts by claiming that conscious intent is not a necessary condition for the order found in an artifact. Is it not possible, it may be asked, that conscious intent is *not* a necessary condition for functionality to be possessed by artifact or part of an artifact? Christopher Boorse,[8] for example, claims that the actual intended result z-ing of an artifact or part of an artifact x by a designer is not a necessary condition for z-ing to be a function of x on the grounds that x may have functions totally unknown to its maker. According to Boorse, there have been mechanisms that achieved their desired goals without being understood by the people who built them. For example, people have known since ancient times that yeast is a necessary ingredient in alcoholic fermentation long before its exact function in this process was known. But

> the actual function in fermentation is to produce enzymes which catalyze the conversion of sugar to carbon dioxide and alcohol. Presumably, then, that has always been the function of yeast in brewing devices. It did not suddenly acquire this function with the advent of chemical theory. But brewers with no knowledge of enzymes cannot intend their yeast to produce them.[9]

Boorse concludes that the concept of functionality in the case of artifacts is independent of the notion of a designed or intended effect.

Perhaps Boorse is correct in asserting that designers' intentions are not a necessary causal condition for the presence of functionality of artifacts on the grounds that artifacts or parts of artifacts may have functions wholly unknown to their makers. But if he is, it is for the wrong reasons. For this example is not a case where a human creation has a function unknown to its designer. The function of yeast in brewing was known to brewers in ancient times. Yeast in brewing has the functional effect of changing certain sorts of foodstuffs to alcoholic beverages. Granted the microscopic mechanisms by which these changes take place were not and perhaps are still not known to many brewers. Nonetheless, the functional effects are known in a general way. It is accordingly unduly restrictive to claim that in these cases the brewer did not know the function of yeast in brewing. Thus, in spite of Boorse's objection, there is little reason to deny that intentions are necessarily a part of functional talk in the case of artifacts.

But certainly not just any natural entity appears to be the result of conscious design. Which things do and which do not? Kant attempts to answer this question in the Analytic of the Teleological Judgment in the Critique of Teleological Judgment by putting forward a criterion which is meant to distinguish things which possess purposiveness from those which do not:

> Experience leads our capacity of judgment upon the concept of an objective and material purposiveness, i.e., upon the concept of a purpose of nature only if a relation of cause to effect is estimated which we comprehend as lawful and only thereby find possible by placing at its basis the idea of the effect of the causality of the cause as the fundamental condition of the possibility of the first (Ak: 366-7; B: 212-3).[10]

This passage suggests that purposiveness arises where there is a causal relation in which the idea of the effect may *prima facie* be cited as an explanation for why the cause of that relation occurred.[11] Thus, to describe nature teleologically is to implicitly offer a description with an explanatory ring. Following this suggestion, "may *prima facie* be cited," or in Kant's words, "we comprehend as lawful," has the following force. If we consider a rise in temperature causing the ice on a pond to crack, it does not seem legitimate to cite the effect, the cracking of the ice, to explain why there is a rise in temperature. In this causal relation and others of its kind, it seems legitimate to explain the occurrence of the cause only by referring to its antecedent conditions. One could not say with propriety, for example, that the temperature rises *in order to* crack the ice on a pond. But consider a different example. Consider the flame cells of a flatworm causing the expulsion of excess water from the flatworm's system. In such a causal relation it seems legitimate to cite the effect, the expulsion of excess water from the flatworm's system, to explain the activity of the flame cells. Thus, Kant offers a linguistic test for distinguishing the teleological from the non-teleological. If it seems legitimate to cite the effect of a causal relation to explain the cause — ignoring for a moment what concept of explanation he is working with — then purposiveness is present in nature.

To what then is Kant attributing purposiveness, i.e., what is it that is purposive? According to Kant, it is the effect in certain unique kinds of causal relations which is the purpose. Thus, it is the effect which is purposive or possesses purposiveness, for it is "the idea of the effect of the causality of the cause" which is "the fundamental condition" of the cause. The effect is to be thought of as the *purpose of the cause* or the *reason* why it occurs. And notice that Kant's attribution of purposiveness to the effect accords with the concept of functionality; it is a certain effect which is a *function*. Later we shall see that among the purposes of nature, Kant includes organismic functions. Kant's preliminary criterion for distinguishing purposive from non-purposive features in nature then is:

> y possesses purposiveness or is a purpose of nature if and only if there is a causal relation, x causes y, for which it is *prima facie* legitimate to cite the idea of y to explain the occurrence of x.[12]

According to Kant, we can determine whether or not something appears to be the result of conscious design by deciding whether or not it seems legitimate to cite the idea of it to explain its cause.

But Kant says that the *idea* of the effect is the "fundamental condition" of the "possibility" of the cause. Is there any special significance to this? Consider the causal relation of the heart's pumping causing the blood in a mammal to be circulated. Now suppose here that it would in fact make sense to explain the occurrence of the heart's activity in terms of the idea of the resultant effect of this activity. What would this supposition entail? Certainly it need not entail that the subsequent circulation of the blood *caused* the prior activity of the heart's beating. For, as Hugh Lehman indicates,

> In a way we might say that circulation of blood causes the activity of the heart, but this is not a case of circulation at one instant causing the activity of the heart which produced that bit of circulation. Circulation of the blood is necessary for the organism to survive and if it survives subsequent heart activity may occur. This is a case of prior circulation determining (or partially determining) subsequent heartbeats.[13]

It might nevertheless be suspected that some sort of etiological claim is being made by this supposition. What *sort* of etiological claim? If the circulation of the blood possesses designedness, it would in fact be correct to explain the heart's activity in terms of the idea of its subsequent effects, or to be more specific, a *designer's idea* of its subsequent effects. In explaining the heart's activity in terms of its effects we would really mean that a prior intention or *idea* that the heart circulates the blood causes the occurrence of the heart's activity.

If organismic activity confirmed the presence of designedness, those attempts to explain the occurrence of this activity in terms of the idea of its effects would be analyzable in the above manner. Such explanatory citations would then be, as R.B. Braithwaite says,[14] "reducible" to straightforward causal explanations. However, the cause of the occurrence of this activity is not a conscious intention. Thus, purported explanations of the occurrence of an event or process in terms of the idea of its effects may not be understood in this way. Yet, if such supposed explanations are not reducible in this manner, the estimation that purposiveness is present in nature is, to say the least, puzzling.

Kant's concept of purposiveness is thus philosophically problematic. Indeed, there are two schools of thought which deny that biological phenomena possess purposiveness. One school, which Kant calles "realism of the purposiveness of nature" (Ak: 72, 391-2; B: 238-9), holds that organismic activity is designed either by an intelligent Being external to this activity or by a vital force acting internally within it (Ak: 72, 391-2; B: 238-9). The former version of realism of purposiveness, "theism" (Ak: 72, 391-2; B: 238-9), holds that organismic activity is goal directed because there is a director, a Supreme Being at work. Likewise, according to theism, the traits of biological organisms get their functions through the conscious design of this Being. The latter version of realism of purposiveness, on the other hand, "hylozoism" (Ak: 72, 391-2; B: 238-9), holds that the goal-directedness and functionality of this activity are due to the presence entelechies acting designedly. The school of realism then, which receives its name because its adherents hold that conscious intentions *are* the source of organismic activity, denies that biological activity merely appears to be the result of conscious design. Rather, both theism and hylozoism assert that organismic activity is consciously designed.

A second school, which Kant calls "idealism of purposiveness" (Ak: 72, 391-2; B: 238-9), denies that biological organisms and organismic activity exhibit purposiveness because there is *no* conscious effort or design involved in the origin of such phenomena. According to this school, Kant's "purposes" are not real but imaginary. There are also, so Kant says, two versions of idealism of purposiveness, atomism and Spinozism (Ak: 72, 391-2; B: 238-9). Both atomism and Spinozism, in holding that organismic activity cannot be accounted for by claiming that it is consciously designed, affirm that purposiveness must be illusory and capable of being explained away.

Both schools of thought share the common presupposition that for purposiveness to be truly predicable of a thing, it must in fact be the result of conscious design. In the case of realism, since conscious design is thought to be present, it is denied that the effect possesses mere purposiveness. In the case of idealism, since there is no conscious intention in the etiology of the effect, it is also denied that the effect possesses purposiveness. But can one claim that a thing possesses purposiveness without being forced to say that it possesses designedness? Clearly an affirmative answer to this question would put one into conflict with the two schools of thought in question.

It is just such a position that Kant maintains in the Critique of Teleological Judgment. Here Kant contends that purposiveness is truly predicable of a thing even though it does not possess designedness, while insisting that the notion of purposiveness is nevertheless a meaningful and useful one. More importantly, we shall see that Kant's contrast between realism and idealism, and his refusal to join either camp, betrays his mechanism *and* his anti-reductionism: he refuses to equate purposiveness and designedness, but he also refuses to dissolve biology into physics. Moreover, we shall see that though Kant's use of the term "idea" in this context is not to be understood as "the intention of an intelligent designer," the meaning of "idea" will be crucial to unraveling his position.

Kant distinguishes between two different sorts of things which may be said to possess purposiveness in the two sentences that immediately follow the passage cited above.

> This [experience leading our capacity of judgment upon the concept of an objective and material purposiveness] can occur in two ways: either we consider the effect immediately as art product or as material for the art of other possible natural beings, thus either as a purpose or as a means to the purposive use of other causes. This latter purposiveness is called usefulness (in the case of man), or also advantageousness (in the case of any other creature), and is merely relative; whereas the former is an inner purposiveness of natural beings (Ak: 63, 367; B: 213).

The idea of the effect may serve as the explanatory ground of "the causality of the cause" when the effect is "material for the art of other possible natural beings" or an "art product." In the former case the effect exhibits relative purposiveness, while in the latter it possesses inner purposiveness.

As we shall see, by "art product" and "natural being," Kant means "living organism." Thus, when Kant considers purposiveness to be a characteristic of a thing, the thing is either a living organism or the material for the use of some living organism. Where there is some sense to explaining the occurrence of an event in terms of the idea of its effect, the effect is either the material for the use of some living thing or a living thing *per se*. In such cases, nature may be said to appear to be the result of conscious design. (Hereafter, unless otherwise indicated, when it is said that the natural world appears to be the result of conscious design, it is meant that relative and inner purposiveness is found in nature.)

On the basis of the foregoing, Kant's criterion of purposiveness may be expanded by saying

> y is a purpose of nature or possesses purposiveness if and only if there is a causal realtion, x causes y, for which it is *prima facie* legitimate to cite the idea of y to explain the occurrence of x in virtue of y being either the material upon which a biological organism subsists or a biological organism *per se*.

In the following sections these two conceptions of purposiveness will be further examined.

3. Relative purposiveness

Kant's distinction between relative and inner purposiveness indicates two different teleological relationships in which biological organisms are found. The former indicates the teleological relationship existing between a living organism and elements *external to* the organism, that is, physical and biological factors both of which are components in ecosystems.[15] The latter points to the teleological relationship existing between a living organism and the various parts internal to it. In the present section the first of these two conceptions of purposiveness will be further explored.

According to Kant, certain things seem to be explicable in terms of the idea of their effects — effects which are advantageous to some living organism. To take one of Kant's examples, consider the withdrawal of the seas in Northern Europe, despositing huge tracts of sand that enabled the growth of extended pine forests in that region. In response to this case Kant asks "whether this primeval desposit of tracts of sand is a purpose of nature for the benefit of the possible pine forest there" (Ak: 63, 367; B: 214)? The tracts of sand are advantageous for the growth of pine trees. But can we say that the desposit of tracts of sand advantageous to the pine trees possesses purposiveness? Can the idea of the effect of the withdrawal of the sea, the deposit of tracts of sand, be *prima facie* cited to explain why the withdrawal of the sea occurred? To his own question Kant answers: "So much is clear that if we assume this as a purpose of nature, we must also admit that sand to be a relative purpose to which the ocean shore and its withdrawal were the means..." (Ak: 63, 367; B? 214). There is here a *prima facie* explanatory sense to a claim like "The sea withdrew for the sake of the pine trees which depend upon the soil conditions created by this withdrawal."[16] In this case, the deposit of sand possesses external purposiveness, or to put it differently, the sand is a relative purpose of nature. This points to the central logical character of the claim that something possesses external purposiveness. It is not the effect *per se* whose idea may be *prima facie* cited to explain the causal efficacy of the cause. Rather the fact that the effect is *used by a living organism* injects these sorts of citations with an apparent explanatory import.

Is every cause which has an effect that is of use to some living organism explicable in terms of its effect? Not according to Kant. Where the effect of a cause, for example, a bird, serves as the material for another being, say man, in that man uses the bird's plumage to adorn his clothes, "we cannot even here assume a relative natural purpose" (Ak: 63, 368; B: 214). But how are we to distinguish cases like the latter from genuine cases of relative purposiveness? Kant answers

> Only *if* we assume that men are to live on the earth then the means must be present without which they could not sustain themselves as animals and even as rational animals (in however low a degree)... (Ak: 63, 368; B: 214-5).

In the case of the bird being used by man, external purposiveness is not present. Even if we assume that men must live on earth as it is now, using a given species of bird for its plumage as adornment is not necessary for the survival of man as a rational being.

The extended tracts of sand in Northern Europe, on the other hand, possess external purposiveness because if we *assume* that the pine forests *must exist* in Northern Europe, then the tracts of sand are indispensable for the pine forests to exist *there*. To take another example, the high energy cosmic radiation coming from beyond the solar system is used by the pine trees as a source of energy. But it cannot, on this basis, be considered a relative purpose of nature. For it does not provide a significant amount of energy to the pine trees, though it is used by them.[17] However, if we assume that pine trees must exist in Northern Europe, the sand may be considered a relative purpose of nature; the sandy soil, according to Kant, is indispensable to the pine trees there. We cannot, of course, legitimately make this assumption, for the tracts of sand could exist in Northern Europe without leading to the development and growth of pine forests.

The foregoing enables us to draw out the following as Kant's conception of external purposiveness.

> Where there is a causal relation, x causes y, y possesses external purposiveness (or y is a relative purpose of nature) if and only if it is *prima facie* legitimate to cite the idea of y to explain the occurrence of x in virtue of y's being indispensable to some living organism at a definite time and place, when it is assumed that this organism must exist at this time and place.

Let us reflect on this conception of external purposiveness for a moment.

One cannot help but notice that Kant's discussion of external purposiveness is inelegant in that he does not make it clear whether the effect in those causal relations in which external purposiveness arises is to be construed as an event or an object. More specifically, he frequently shifts back and forth between referring to the effect as if it were an event and an object. For example, he talks about the effect as *the sea leaving behind tracts of sand* and as *the tracts of sand*. Which, then, is to be considered as the relative purpose of nature, the event or the object?

In spite of the fact that Kant sometimes talks about the effect in both ways it could not be the effect *qua* depositing of sand which is the relative purpose of nature. For an event is not the sort of thing that serves as the raw material upon which a living organism could subsist. It is reasonable to think then that by effect Kant must have in mind some sort of object or stuff that is of use to a living thing. The preceding is also borne out once it is noticed that Kant's concept of external purposiveness appears to be an attempt to partially explicate our concept of an ecological relationship. And ecosystems are for the most part described as being composed of objects. In the example of the pine trees and the soil conditions allowing for their growth, Kant seems to be trying to isolate the nature of the relationship which exists between the elements of an ecosytem. Kant's other examples also indicate that it is the teleological character of the relationship between biological organisms and their environment which is his concern. As Kant says

> So too, if cattle, sheep, horses, etc. are to exist in the world, then there *must* be grass on earth, but saline plants must also grow in the desert if camels are to thrive; and again these and other herbivorous animals must be met with in numbers if there are to be wolves, tigers, and lions (Ak: 63, 368; B: 214; my emphasis).

Thus, though Kant sometimes speaks of the effect as an event, he thinks that it is an *object* which is a relative or external purpose of nature.

But in the *Prolegomena* Kant claims that "every effect is an event or something that happens in time..." (Ak: 53, 343; BK: 91). That is to say, according to Kant (as well as many other philosophers), only events enter into causal relations. Is there an inconsistency here? As we shall see, this shift back and forth between event and object also plagues Kant's discussion of internal purposiveness. However, Kant probably has a tendency to use event-language in his discussion of the teleological character of ecosystems because his concern is with the explanatory content of descriptions of ecological relations. And Kant doubtless believes that explanatory talk can only occur in the context of causal talk. Once we distinguish between causation and explanation, and realize that Kant's concerns are explanatory, then there may not be an inconsistency in his regarding objects as entering into causal relations. He does so because his object of study is the teleological judgment (which is intrinsically explanatory), not because he thinks that objects do in fact enter into causal relations.

A second but related difficulty with Kant's notion of relative purposiveness is connected with its entailing that a biological or physical factor is indispensable to a living thing. It is this indispensibility that contributes to the teleological character of ecosystems. Now it is important to observe that present-day biologists often claim that certain things are indispensable to biological organisms in ecosystems. For example

> Practically all of the energy in an ecosystem originates as radiation from the sun, that is, *solar radiation*. Minor amounts of high energy cosmic radiation coming from beyond the solar system also enter our environments.... Similar high energy radiation is received in certain local environments due to the presence of radioactive rocks or fallout. Around volcanoes and hot springs, some geothermal energy (heat) of non-solar origin is added to local environments. But all these latter sources are relatively minor; it is solar radiation that provides the necessary energy to heat the environment and to drive the ecosystems by means of energy storage at photosyntheses.[18]

> Although it primarily consists of relatively inert nitrogen, the atmosphere supplies organisms with the necessary oxygen for respiration and with carbon dioxide and water vapor required for synthesis.[19]

But whereas it is Kant's claim that in ecological relationships a physical or biological factor (the sandy soil, the saline plants) is indispensable to a biological organism (the pine tree, the camel), the passages above assert that a certain ecological factor (solar radiation, atmosphere) has *effects* (providing energy, providing oxygen) which are indispensable to the biological elements of the ecosystem. Thus, as far as contemporary biologists are concerned, Kant is mistaken in claiming that it is the *material* and not its effects which are indispensable to a biological organism.[20]

The question of whether it is the component or the effect of a component in an ecosystem which is indispensable to the biological organism might bring to mind Hempel and Nagel's analysis of function statements.[21] Both Hempel and Nagel see function statements as assertions that something is necessary for the organism as a whole. But whereas Hempel claims that to ascribe a function to a trait of a biological organism is to say that the trait has effects which satisfy a condition necessary for adequate functioning of the organism, Nagel asserts that to ascribe a function to a trait of a biological organism is to in part assert that the trait itself is a necessary condition for some effect or for the whole organism.[22] More specifically, Hempel would construe the functional ascription "the function of the kidneys in mammals is to eliminate waste" as the claim that kidneys have the effect of eliminating waste, an effect which is indispensable for the proper functioning of mammals. Nagel, on the other hand, would construe the function ascription in question as asserting that the kidneys are a necessary condition for the elimination of waste in mammals. Yet certainly the latter is factually mistaken. There are, after all, kidney machines. Nagel recognizes this and suggests that we could qualify his analysis by considering it as referring to *normal* mammals.[23] Nevertheless, of the two suggested analyses, Hempel's is more accurate. For there are functional parts of normal mammals which are not necessary for the survival of the mammal as a whole, e.g., the human "little finger," even though they have effects which are necessary for the proper functioning of the mammal.

Strictly speaking then, it is not the component within the eco-system which is necessary for a given biological organism in that system.[24] Rather, if anything is indispensable, it is *the effects* of the component. The sandy soil is not necessary for the pine trees; the pine trees could get along quite well without the sandy soil if the effects of the sandy soil could be provided by something else. Whether any item external to a biological organism is indispensable to it seems doubtful then. This in turn makes it doubtful that there ever was or will in fact be an instance of Kant's concept of external purposiveness. Hence, construed as an attempt to conceptually grasp the relationship between a biological organism and its external conditions, Kant's concept of external purposiveness is inaccurate, though not totally misguided.

4. Internal purposiveness

Most of Kant's discussion in the Critique of Teleological Judgment is directed toward his conception of internal purposiveness. Moreover, as we shall see, Kant's anti-mechanistic remarks apply to cases of internal purposiveness only. On the basis of what has been said so far, the following may be considered as Kant's preliminary construal of internal purposiveness.

y possesses internal purposiveness if and only if there is a causal relation, x causes y, for which it is *prima facie* legitimate to cite the idea of y to explain the occurrence of x in virtue of y's being a biological organism.

Kant's conception of internal purposiveness reflects what he considers to be the content of the judgment that something is a living organism. Although more will be said in detail about the Kantian inquiry in the following Chapter, it is important to realize here that Kant seeks to uncover what is entailed by our assertion that nature appears to be the result of conscious design. He believes, as already indicated, that this assertion applies either to ecological relationships or living things. As we saw in the previous section, the former application is ultimately traced by Kant to the claim that some occurrence in nature produces an element which is indispensable to a living thing. Kant's concept of internal purposiveness, then, is an attempt to fill out another respect in which nature is judged to appear to be the result of conscious design.

In his explication of the second concept of purposiveness, Kant points to what he believes to be the key features of the relationship between a biological organism and the parts internal to it by contrasting artifacts with "natural purposes," his expression for biological organisms. This contrast is essentially one between organized things and self-organizing things. For a natural purpose is "an organized and self-organizing being" (Ak: 65, 374; B: 220), whereas an artifact is but an organized thing. More specifically, Kant believes there are two criteria which, taken jointly, are necessary and sufficient conditions for something to be a natural purpose. One of these conditions, however, is considered by Kant to be necessary and sufficient for a thing to be an artifact.

Although Kant's present goal seems straightforward enough, as we shall see, his description of a natural purpose comes close to being hopelessly obscure. But it is not *philosophically* obscure. Nevertheless, we shall see that Kant's contrast between living things and artifacts nearly thwarts his attempt to analyze the estimation that internal purposiveness is present in nature.

The fact that Kant's concern lies with "our teleological judgments about nature" makes it somewhat easier to understand his initial attempt to isolate the first feature of a natural purpose (Ak: 73, 392; B: 239). That is, Kant, knowingly or unknowingly, makes two attempts to characterize the first feature of a living thing, a feature which he says is also shared by artifacts. His first effort emerges in the following passages.

> For a thing to be a natural purpose it is, in the first place, necessary that the parts (with respect to both their presence [*Dasein*] and form) are possible only through their reference to the whole. For the thing itself is a purpose, and so it follows that it is comprehended under *a concept or an idea* that *must determine a priori everything which is to be contained in it* (Ak: 65, 373; B: 219; my emphasis).

> But if a thing as a natural product is ... to be possible only as a natural purpose and without the causality of the concepts of rational beings external to it, then it is, in the second place, necessary that its parts should so combine in the unity of the whole that they are reciprocally cause and effect of each other's form. For in such a way as this it is alone possible that *the idea of the whole can conversely (reciprocally) in turn determine the form and combination* [*Verbindung*] *of all the parts* (Ak: 65, 373; B: 220-1; my emphasis).

How are we to understand Kant' claim that "the idea of the whole" must "determine the form and combination of all the parts?" Kant maintains that in causal relations in which internal purposiveness arises, the idea of the effect, the whole organism, determines not only the "form," i.e., the *structure* of the parts, but also the arrangement in which the parts are found. Moreover, the idea of the whole "must determine *a priori* everything which is contained in it." But this first criterion is supposed to be applicable to instances of purposiveness. Thus, though in cases of purposiveness, the idea of the whole determines the structure and arrangement of each part, it does not do so "as cause" (Ak: 65, 373; B: 221), otherwise designedness would be present. What then does Kant mean here by "an idea of the whole determining the parts?"

Since Kant wishes to analyze a kind of teleological judgment, we must assume that by "idea" he means a concept of a certain sort. Thus, when matter is estimated to be living, it is the *judging subject's* concept of the whole which "determines" the parts. But though it is very Kantian to claim that concepts determine particulars in a quasi-ontological sense (e.g., the categories "determine" the central features of nature as an object of experience), and that, generally speaking, the epistemic and the ontic are not so easily separated in Kant, the sense in which the concept or idea of the whole "determines" the part here is entirely epistemological. For consider, to take one of Kant's (and one of the eighteenth century's favorite) examples, a watch. It might be argued that the idea, or better, the plan of its designer, can generate a description of all the functional effects of the parts of the watch. We could in principle study such an idea, where an idea is construed as a plan, and discover all those *functional* characteristics of the watch. In this way these functions could be known wholly independently of experience. Thus, in judging that something is a natural purpose, the subject's idea of the whole living system determines the form and combination of the parts in the sense that this concept can generate these functional descriptions.[25]

But it is hard to see how the fact — if it is a fact — that the concept of a whole entails a description of all of its parts is relevant to the judgment that something is alive. We could just as easily say that the concept of a magnetic field entails descriptons of all of the parts of the magnetic field. Yet this would hardly be a very telling consideration for classifying it among vital phenomena.

However, we shall see that though the claim that the idea of the whole entails a description of all of its parts does not uniquely distinguish the teleological judgment, a certain type of use (a *reflective* use) of a particular idea (the *idea of design*) will do just this. Moreover, this initial attempt to express the first distinguishing mark of a natural purpose does exhibit one thing about Kant's conception of internal purposiveness. In those causal relations in which internal purposiveness arises, the effect is the whole organism and the cause one of its parts which in some way serves to maintain the organism. The effect is a natural purpose; the cause or part is a means. As we shall see, *the way* in which part is related to whole will be of central importance to Kant's conception of organismic activity.

But this points to the oddity which Kant's conception of internal purposiveness shares with his conception of relative purposiveness. He speaks of the former arising in causal relations, but characterizes the cause and effect of these relations as objects, the cause being an organ and the effect the whole living thing. If, however, only events can enter into causal relations, by "part" he should mean "a biological event" and by "whole" that "collection of events which is taken to be a living system." Now there are certainly times when Kant suggests that by "part" and "whole" he means just this. But his leaving "part" and "whole" ambiguous in this way detracts from his explication of this second kind of teleological judgment.

We must now consider Kant's second attempt to state the first distinguishing characteristic of a living thing, a trait which makes natural purposes, like artifacts, organized beings. According to Kant,

> In such a product of nature, every part is thought of as existing only *by means of* the rest and also *for the sake of* the others and the whole, that is, as an instrument (organic): but this is not enough (for it could also be an artificial instrument and thus only represented to be possible in general as a purpose) (Ak: 65, 373-4; B: 220).

This feature of causal relations in which internal purposiveness arises points to the functionality of the parts of a natural purpose. Where internal purposiveness is possessed by an effect, the cause is functional. Thus, according to Kant, to claim that

> part p has the function e in system S

is just to say

> part p exists in system S for the sake of other parts $p_1...p_n$ in S and S *per se*.

Notice that this tells us nothing about what it is for a part of a natural purpose to have a function. In fact, if one glances over the literature on functional analysis, it becomes clear that the claim that the parts of a natural purpose exist for the sake of others and the whole does not go beyond the claim that these parts are functional. Achinstein points out that analyses of function statements fall into three general categories.[26] The first is what he calls "the good-consequence doctrine" according to which doing y is x's function only if doing y confers some good.[27] The second, which he calls "the goal doctrine," asserts that y is a function of x if and only if x does y in S and doing y contributes to some goal which x or S has.[28] The third position Achinstein calls "the explanation doctrine."[29] This is the view that function statements provide etiological explanations of the existence or presence of the functional items. Claiming that x does y for the sake of S and other parts in S is not only compatible with these three sorts of analyses, but could also serve as the analysandum of any of these sorts of analyses.

Of course, it is possible to go back to Kant's account of external purposiveness, draw from it the analysis of functionality found there, and apply it to his account of internal purposiveness. If this is done, · the claim

> part p exists in system S for the sake of other parts $p_1...p_n$ in S and S *per se*

will mean nothing more than

> part p is indispensable to other parts $p_1...p_n$ in system S and S *per se*.

The result would be, of course, an account of function statements like that of Nagel. But as we shall see in the final Chapter, Kant is committed to no particular analysis of functional ascriptions to parts of living things, as long as the analysis provided is not given in terms of some designer's intentions.

Later in the Critique of Teleological Judgment this functionality requirement gets expressed in the form of the teleological maxim of the reflective capacity of judgment.

> *An organized product of nature is one in which every part is reciprocally purpose and means* (Ak: 66, 376; B: 222).

I shall discuss the functionality requirement and its form in Chapter V. But at this point it is important to realize that Kant's inclusion of the functionality requirement in his definition of a natural purpose is a mistake. For, although it is true that Kant accepts this requirement in some form or other, it is not part of the content of the estimation that purposiveness is present. The teleological maxim is rather a part of Kant's attempt to show how teleological judgments are "possible." Although I shall discuss in detail Kant's Critical methodology in the following section, suffice it to say here that by including this functionality requirement in his definition, Kant is conflating his "analysans" (the teleological maxim) with his "analysandum" (the

24

estimation that internal purposiveness is present in natural). It is the latter which he is supposed to be laying out here.[30]

Still, Kant's contrast between designed things and undesigned things which are "self-organizing" is relevant. For the second feature exhibited by those causal relations in which internal purposiveness arises is lacking in those which confirm the presence of designedness. It is, moreover, the condition which will be central later. This distinguishing mark of a natural purpose is "that its parts thereby combine in the unity of a whole in such a way that they are reciprocally cause and effect of each other's form" (Ak: 65, 373; B: 219-20). According to Kant, a "natural purpose must bear itself alternately as cause and effect. This, however, is a somewhat inexact and indeterminate expression which needs derivation from a determinate concept" (Ak: 65, 372; B: 218). A living thing for Kant is both a cause and an effect. Kant is right. This is an obscure remark. But he does attempt to unpack it.

A tree produces itself as an *individual*. To be sure, this kind of effect we call growth; but this growth is altogether different from any other increase according to mechanical laws and is to be regarded as generation.... The matter that it adds to itself is previously processed by this growth into a specifically peculiar quality which the mechanism of nature external to it cannot produce, and develops itself by means of a material so that its composition is its own product. For although with respect to the constituents that it receives from nature external to it, it can only be seen as an educt. But in the separation and recombination of this raw material we see such an originality in the separating and formative faculty of this kind of natural being that is infinitely beyond the reach of art (Ak: 64, 371; B: 217-8).

In a watch, one part is the instrument for moving the other parts, but the wheel is not the effective cause of the production of the others; no doubt one part exists for the sake of others, but it does not exist by their means.... Hence a watch wheel does not produce other wheels; still less does one watch produce other watches, utilizing (organizing) foreign material for that purpose. Hence it does not replace by itself parts of which it has been deprived, nor does it make good what is lacking in a first formation by the addition of the missing parts, nor if it has gone out of order does it repair itself — all of which, on the contrary, we may expect from organized nature. An organized being [a living thing] then is not a mere machine, for that has merely *moving* power, but it possesses in itself *formative* power and to be sure such a formative power that is communicated to its materials which lack it (they are organized); therefore they have a self-propagating formative power... (Ak: 65, 374; B: 220-1).

A living thing is not only "the effect of the concurrent moving powers of the parts" (Ak: 77, 407; B: 257); we also conceive of its "parts... as dependent on that of the whole" (Ak: 77, 407; B: 256). In a living system then part and whole are uniquely related. Between whole and part exists not only a causal relation, but a formative relation.[31]

In what respect then does the part of a living thing cause the whole? Although this may sound as though Kant is smuggling in the notion of an Aristotelian formal cause, saying only that the parts *comprise* the whole, he is, no doubt, talking about the parts as being efficient causes of the whole. We can say, for example, that cellular processes within the bone marrow cause the development of white blood cells, and insofar as the latter in part comprise the whole, this gives credence to speaking of the cellular processes as a cause of the whole. Thus, the causality of the part here is effective.

But a living thing is likewise a whole which forms, i.e., creates the parts. This general statement no doubt reflects the linguistic propriety of the assertion, for example, that something like a starfish (a biological whole) often regenerates one of its lost or damaged limbs (parts). More specifically, in claiming that the part of a living thing is formed by the whole, Kant is pointing to the fact that the part of a living thing is the effect of its generative and self-directing powers. Kant, that is, is pointing to a feature of living systems which William Harvey, the pioneering physiologist, had emphasized over a hundred years earlier: "they derive their origin from a certain primary something or primordium which contains within itself both the 'matter' and the 'efficient cause'; and so is, in fact, the matter out of which, and that by which, whatsoever is made."[32] Indeed, Kant even speaks with approval of the theory of "epigenesis," Harvey's term for this view of living systems. Thus, the capacity to organize and create itself by bringing together and modifying matter — the capacity which Kant is referring to when he speaks of living things as forming their parts — might be referred to as the "epigenetic" capacity of living things. This capacity makes living things stand apart from inorganic nature and, as we shall see, will lead Kant to deny that living phenomena are "reducible" (in one sense of the word) to highly sophisticated physical processes.

One might be tempted to respond to Kant's discussion of the epigenetic capacity of living things by saying:

> Suppose that a mechanism is constructed with the following characteristics: It is a servo-mechanical device which is goal directed with respect to a certain type of radiation — i.e., when its sensory units detect such radiation, it seeks out their source (somewhat like an acoustical homing torpedo). These sources are energy sources, and this device has been equipped with mechanisms which allow it to replenish its internal energy store from sources emitting this type of radiation. Its internal energy store can be destroyed by overcharging, and it charges at a rate proportional (among other things) to the strength of the source. It has also been provided with a time-limiting circuit, which causes it to break off contact with a source after a definite amount of time (say t seconds), and to resume its search for sources, ignoring the last source contacted. The behaviour of this device in an environment containing sources of the appropriate type which are of varying strength will be as follows: it will wander through the environment, from source to source, breaking off contact with each source after t seconds until it encounters some source greater than a given strength, v. At such a source, its power supply will be destroyed through overcharging in less than t seconds, and it will stop, since the destruction of its internal power source renders it incapable of continuing the search procedure.[33]

"Why is this device not a natural purpose?", it may then be asked. Strictly speaking, of course, such a device is not a natural purpose because it possesses designedness. By definition, anything whose etiology includes conscious intentions possesses designedness, not purposiveness. But even so, this mechanism does not possess an "epigenetic" capacity in Kant's sense of the term. For according to

Kant, a natural purpose "produces itself generically" (Ak: 65, 371; B: 217). And once produced, a living thing "builds," and not only maintains itself; e.g., an acorn sprout "builds" itself into an oak tree. Kant, it seems, is putting his finger on one of the essential attributes of the living: its capacity to replicate itself. Indeed, in considering the origin of life, we begin considering that point in time when molecules which possess the property of being able to create copies of themselves appear. Being able to create "copies" of itself, as well as possessing the capacity to replace some of its lost or damaged parts, is a mark of the living. It is this sort of characteristic, in addition to the etiological difference, that distinguishes natural purposes from artifacts. Thus, the teleological character of living things — the epigenetic capacities of their parts — provides the distinguishing mark between living and inorganic phenomena, and living things and artifacts. We might say then that a living thing manifests a *designer-like* and *designed-like* aspect in virtue of being a whole which forms its parts.

We can see then that Kant is here offering a suggestion as to why we *cannot* say with propriety that the temperature rises in order to crack the ice on a pond, but *can* say, at even the molecular level, something like "the function of loop I of the tRNA molecule is to bind the specific activating enzyme, the amino acyl synthetase molecule to the tRNA molecule." The legitimacy of the latter remark resides in the fact that the tRNA molecule is thought to be a part in a whole which forms itself, whereas the rise in temperature is not. In Chapter V we will consider Kant's reasons for thinking that *this* difference explains the appropriateness of the functional ascription to the tRNA molecule and the inappropriateness of functionality to the rise in temperature.

It is interesting that in the First Introduction to the *Critique of Judgment*, Kant claims that

> Nature works *mechanically* in view of its products as aggregates, as *mere nature*. But in view of its products as systems, for example, crystallization, all sorts of flower forms, or the inner structure of plants and animals, nature works *technically*, that is, at the same time as *art* (Ak: VI, 217; H: 22).[34]

Notice that here Kant includes crystals along with vital phenomena as natural objects which behave "technically." This suggests the objection that there are inorganic objects, namely crystals, which are wholes that form their parts. For during the addition of parts to a crystalline whole the correct orientation of these parts is established by the presence of similar structures already formed. If this is true, so it might be argued, a crystal is an instance of a natural purpose.

The problem is, however, that in the structural growth of a crystalline solid, a growth the requires the same pattern to be repeated in three dimensions by a regular arrangement of atoms from center to surface, the interior of the structure is inaccessible.[35] And insofar as it is inaccessible, it might be claimed that it has no function, in which case a crystal would seem to be a thing different in kind from a natural purpose. Regardless of which side we favor, the analogy between living things and crystals has in the past been,[36] and continues to be invoked to explain the shape, growth, and reproduction of living systems.[37] To some the phenomenon of crystallization suggests that there is not a difference *in kind* between the organic and inorganic world, but only *in complexity*. To others, and I suppose the Kant of the Critique of Teleological Judgment is one, the formation of a crystal is the closest a causal relation in nature can come without indicating the presence of internal purposiveness. The question is "Why?" The answer is the answer Kant would give to those who wonder why some sophisticated mechanical devices are not natural purposes, viz., crystals (and ecosystems for that matter) are *not* epigenetic wholes. They do not instantiate the same kind of purposiveness that living systems instantiate. It is true that crystals and ecosystems "grow," in Kant's sense of the term. They add matter to themselves and develop themselves "by means of a material so that" their "composition" is their "own product" (Ak: 64, 371; B: 217). And ecosystems are to some extent self-repairing. (Can a crystal repair itself?) However, neither can replicate themselves; neither one, that is, "produces itself generically" (Ak: 64, 371; B: 217).

In any case, we can now fill out Kant's conception of internal purposiveness in its entirety.

> Where there is a causal relation, x causes y, y possesses internal purposiveness (or y is a natural purpose) if and only if it is *prima facie* legitimate to cite the idea of y to explain the occurrence of x in virtue of (a) x being a part within y, (b) x being the cause of y, and (c) y forming x.

This and his concept of relative purposiveness are best viewed as characterizations of the relations between a biological organism and its environment, and a biological organism and its parts. The numerous instances of both of these concepts stand out among the vicissitudes of the natural world as appearing to be the result of conscious design. In the remainder of this study we will investigate the implications which Kant draws from these two conceptions via his Critical approach, an approach we will now examine.

Notes

1. R.A.C. Macmillan, in *The Crowning Phase of the Critical Philosophy* (London: Macmillan, 1912), p. 270, claims: "In the *Critique of Judgment*... a subtle change enters into Kant's view of the origin of species. He... advances towards the view that matter is brought into existence by the same creative act as endows it with life; or, to take it from the other side, that matter is a divine creation and may therefore contain in its orginal constitution the purposive combinations of organic life."

2. For example, H.W. Cassirer, in *A Commentary on Kant's Critique of Judgment* (New York: Barnes & Noble, Inc., 1938), p. 311 states: "Kant's problem is: How is it possible to regard certain objects as purposes of nature? It seems to be impossible to do anything of the kind, for nature as an object of experience... does not give us any indication that it obeys yet another principle, that it produces objects purposely." Likewise, S. Körner claims, in *Kant* (New York: Penguin Books, 1977), pp. 197-8: "It therefore becomes an important task of the critical philosophy to examine the notion of purpose and the manner of its legitimate and illegitimate employment in science." From his opening remarks J.D. McFarland, in *Kant's Concept of Teleology* (Edinburgh: University of Edinburgh Press, 1970), p. v maintains: "The problem of teleology is the problem of whether or not concepts such as 'purpose', 'end', 'design', have any value for the investigation and explanation of nature. In this study I have attempted to explain why this problem arose for Kant...."

3. This mechanistic trend is spelled out in more detail in Chapter III, Section 1.

4. Two recent books, Gordon G. Brittan, Jrs.'s *Kant's Theory of Science* (Princeton: Princeton University Press, 1978) and W.H. Walsh's *Kant's Criticism of Metaphysics* (Chicago: The University of Chicago Press, 1975), have expounded on the anti-reductionist elements in the Critical Philosophy, the latter implicitly and the former explicitly. Walsh, for example, finds Kant sometimes toying with the view that "sensation must be conceived of as a form of experience which is *sui generis*..." (p. 95), and does not convey immediate knowledge. If this is the case, Kant cannot be seen as believing in the possibility of a phenomenalist reduction of material object propositions to propositions about "the contemplation of something which is private..." (p. 94). Brittan sees Kant's classification of mathematical judgments as synthetic *a priori* as a refusal to acquiesce to the Leibniz-Frege-Russell program of reducing mathematics to logic (Chapter 2).

5. This is my terminology, not Kant's.

6. Kant's term for purposiveness is *"Zweckmässigkeit."* "Purposiveness" in Kant is actually a generic term of which the "purposiveness" in organic nature is a species. When speaking of "purposiveness" generically, Kant means "all order in the world" which

seems "as if it has originated in the purpose of a supreme reason" (A 686/B 714). In describing organic nature and the cause of its order, Kant refers to the *objective material* purposiveness in nature in order to distinguish this sort of purposiveness from the *subjective formal* purposiveness of the aethetic state of mind and the *objective formal* purposiveness of mathematics. I shall not examine the latter type of purposiveness, but will briefly discuss the former in Chapter II, Section 3. When I use the term "purposiveness" I will be referring to objective material purposiveness. Thus, when I show later in this Chapter that Kant's theory of the teleological judgment includes the distinction between functions and goals of living processes, I should *not* be understood as *identifying* biological function with functionality in general and the goal-directedness of living things (their self-regulating, regenerative, and generative processes) with goals in general in Kant. I will – to repeat – be restricting my discussion to concepts of *biological* function and goals as they are used in the Critical texts.

7. Kant does not use the term designedness, though he does speak of causes "working according to design" (Ak: 75, 398; B: 245). I will adopt the term "designedness" as a grammatical analogue to "purposiveness."
8. Christopher Boorse, "Wright on Functions," *The Philosophical Review*, 84 (1976), p. 73.
9. Ibid.
10. There are two remarks to make about this passage, the first of which concerns the phrase "capacity of judgment." The word used by Kant, *"Urteilskraft,"* is most accurately rendered capacity, power, or faculty of judgment. And while it may be a bit more clumsy to render *"Urteilskraft"* as "capacity of judgment" rather than as "judgment," as most translators do, to not do so might lead one to think that Kant has in mind the product of judgment – a linguistic entity – when using this term. Moreover, Kant uses the word *"Urteil"* when he wishes to speak of the product of the capacity of judgment. The ambiguity of "judgment" can thus be avoided by rendering *"Urteilskraft"* and "capacity of judgment." (I will, however, continue to translate *"Kritik der teleologischen Urteilskraft"* as the "Critique of Teleological Judgment.") It might be objected that by rendering *"Urteilskraft"* as "capacity of judgment," I make Kant sound like he is doing a priori psychology. Indeed this translation, so it seems, plays into the hands of Kant's faculty talk. I am sympathetic with this objection. But it must be remembered that *Kant* uses this language. Granted, it is a challenge for any philosophical interpretation of Kant to read his work in a non-psychological way. Still, it violates the "rules of the game" if we *translate away* the psychological overtones of the Critical Philosophy.

My second remark, which concerns the phrase "estimate," is not unrelated to my first. The German word in this context, *"beurteilen,"* is most accurately rendered "to form an opinion of the nature, character, or quality of a thing," and "estimate," in at least one of its uses, captures this sense. *"Beurteilen"* is to be contrasted with *"urteilen,"* the latter term being used by Kant when he is talking about judging things objectively as real in accordance with the categories of the understanding, and "judge" captures *"urteilen"* accurately. The *beurteilen/urteilen* distinction runs parallel to the purposiveness/designedness distinction; our power of judgment enables us to estimate that something possesses purposiveness and judge that a different thing possesses designedness. The word *"beurteilen"* is used, like the word *"Zweckmässigkeit,"* to mark our caution about our knowledge of the origin of certain entities. I will, in characterizing Kant's view, occasionally speak about "judging" things as appearing to be the result of conscious design simply for aesthetic reasons. But the reader, I hope, will be able to tell that on these occasions I am characterizing "the estimation that something." In the course of this work we will, in addition, see that *beurteilen* and *urteilen* are each associated with their own unique concept of explanation.

11. Also see #10 in the Critique of Aesthetic Judgment.
12. In attempting to capture Kant's conception of purposiveness, I have and will continue to assume that it is definitional, and thus use the "if and only if."
13. Hugh Lehman, "Fucntional Explanation in Biology," *Philosophy of Science*, Vol. 32 (1965), p. 16.
14. R.B. Braithwaite, *Scientific Explanation* (Cambridge: Cambridge University Press, 1968), p. 325.
15. It is thus not surprising to find Kant throughout much of the Critique of Teleological Judgment referring to relative purposiveness as external purposiveness.
16. The sense of "explain" here is discussed in Chapter V.
17. W.D. Billings, *Plants, Man, and Ecosystem* (Belmont, California: Wadsworth Publishing Company, 1970), p. 10.
18. Ibid., p. 10.
19. Ibid., p. 23.
20. It is true that in some of his examples (the use of the bird's plumage) Kant clearly is considering whether the *effects* of a material (the plumage of a bird) are indispensible to a living thing (man). However, his general statement of objective material purposiveness together with his initial distinction between internal and relative purposiveness makes it clear that a relative purpose of nature is "the material and not the art of other possible natural beings" (Ak: 63, 367; B: 213), and not its effects.
21. Carl G. Hempel, "The Logic of Functional Analysis," in *Aspects of Scientific Explanation and other Essays in the Philosophy of Science* (New York: The Free Press, 1965), pp. 297-330; Ernest Nagel, *The Structure of Science* (New York: Harcourt, Brace and World, Inc., 1961), Chapter 12.
22. See Lehman, "Functional Explanations in Biology," for a discussion of just this point of difference between Hempel and Nagel's analysis of function statements. It is also to be noted that in Chapter 14 of *The Structure of Science*, Nagel offers an interpretation of function statements which is essentially that of Hempel.
23. Nagel, *The Structure of Science*, p. 292.
24. In leaving this issue I am not suggesting that I regard Hempel's analysis of function statements to be correct. Yet his analysis undoubtedly seems most plausible when it is tested against functional ascriptions to components of ecosystems.
25. In the *Prolegomena* Kant says, "As in the structure of an organized body, the end of each member can only be deduced from the full conception of the whole" (Ak: Intro., 263; Bk: 11).
26. Peter Achinstein, "Function Statements," *Philosophy of Science*, 44 (1977), pp. 341-67.
27. Ibid., p. 342.
28. Ibid.
29. Ibid., p. 344.
30. Again, these matters will be discussed in fuller detail in the following chapter.
31. Following Kant, I will speak of the whole as *forming* its parts and characterize the part as *causing* the whole in order to keep these two sorts of relations between part and whole distinct.
32. William Harvey, *The Works of William Harvey* (in one volume), R. Willis, trans. (London: Sydendam Society, 1847), p. 554.
33. William C. Wimsatt, "Teleology and the Logical Structure of Function Statements," *Studies in History and Philosophy of Science*, 3 (1972), pp. 20-1.
34. Kant wrote two introductions to the *Critique of Judgment*, the first of which was discarded because he thought it too lengthy. In a letter to Beck on 4 December 1792, Kant explains that the First Introduction was discarded "*für den Text unproportionirten Weitläufigkeit*" (Ak: XI, 396).

32

35. François Jacob, *The Logic of Life*, Betty E. Spillman, trans. (New York: Vintage Books, 1976), p. 303 discusses this point of difference between living things and crystals.
36. Again see Jacob, *The Logic of Life*, pp. 254, 282-2, 286, and 303-4 for the history behind this analogy.
37. P. Jonathan, G. Butler and Aaron Klug, "The Assembly of a Virus," *Scientific American*, 239 (1978), pp. 62-9 explain the assembly of a viral particle by analogy to the formation of a crystal.

CHAPTER II

The Kantian endeavor

> it is impossible to find the principle of *a true unity* in matter alone
> in that which is only passive...
>
> Leibniz

1. The Critical methodology

The *Prolegomena* contains Kant's most explicit statement of what the Critical Philosophy is supposed to accomplish theoretically. In the first sentence of the *Prolegomena* we find Kant saying

> These *Prolegomena* are for the use... of future teachers, and even the latter should not expect that they will be serviceable for the systematic exposition of a ready-made science, but merely for the discovery of the science itself (Ak: Intro., 255; Bk: 3).

The science in question here is metaphysics, which he later says

> cannot exist unless the demands here stated on which its possibility depends be satisfied; and, as this has never been done, there is, as yet, no such thing as metaphysics (Ak: Intro., 257; Bk: 4-5).

These statements indicate what perhaps Kant thinks is the most important theoretical goal of the Critical Philosophy; they show that Kant wishes to put forward a set of criteria of adequacy which must be met for a proposition to be classified as "metaphysical." Kant's goal then is to demarcate metaphysics from both science and common sense. We might say that Kant's central aim is meta-meta-physical (his term is "transcendental"); he wishes to establish, to put it somewhat vaguely, the basis of metaphysics.[1] Much of the first *Critique*, as well as the *Prolegomena*, contains Kant's attempt to formulate this.

The main outline of this endeavor is familiar. According to Kant, Hume sets the stage for determining the criteria of adequacy for metaphysics by claiming that there cannot be any such thing as metaphysics at all, if metaphysics is construed as reasoning a priori in a non-analytic way,[2] that is, as consisting of synthetic judgments which are necessary and unviersal.[3] Hume's demand is, of course, founded on the dictum that there can be no idea without a prior sense impression which generates that idea, which is Hume's variation of the empiricist tenet: experience alone is the source of all knowledge. All judgments about matters of fact are formulated on the basis of experience either directly, as in the case of "The table is hard," or in a Pickwickian way, as in "The heart causes the blood to be circulated," where cause is understood to mean "necessary connection." Thus, on Hume's construal, metaphysics is impossible. For experience does not provide the required evidence for the formulation of judgments which are necessary and universal. If the criterion of adequacy for metaphysics is that its propositions are synthetic a priori, then metaphysical reasoning cannot occur.

It is clear that Kant took Hume to be asserting that certain concepts, such as causality and substance, cannot be thought by reason a priori, and consequently possess no "inner truth, independent of all experience" (Ak: Intro., 259; Bk: 7). Kant's famous reaction to Hume — which betrayed his intellectual debt to the rationalists — was that such concepts are "not derived from experience, as Hume had attempted to derive them, but spring from the pure understanding" (Ak: Intro., 260; Bk: 8). They thus can be isolated independently of experience. Nevertheless Kant believed that they have a use restricted to mere objects of experience. Kant *accepted* the remainder of empiricism. Since all knowledge ranges within the limits of experience, a priori reasoning is an ingredient of human knowledge. With this Kant's meta-metaphysical program began to emerge. Metaphysics, for Kant, is reasoning a priori but with a different twist. It consists of a priori reasoning about objects *within* the sphere of experience, and thus of knowledge. Specifying the guidelines in terms of which this a priori reasoning may proceed is tantamount to a statement of the criteria of adequacy which must be met for a proposition to be considered metaphysical.

The transcendental inquiry which sets the guidelines for metaphysical reasoning deals in large part with "the categories of the understanding," most of which comprise the collection of a priori notions in question. The meta-metaphysical enterprise involves stating what these categories are, explaining why they apply to experience, and showing how it is that they apply to experience. These three tasks correspond nicely to the "Metaphysical Deduction of the Categories," the "Transcendental Deduction of the Categories," and the "Schematism of the Categories." But it seems to me that nearly all of what Kant calls the "Transcendental Analytic" in the first *Critique* is meta-metaphysical in nature. The famed Second Analogy, for example, Kant's attempt to establish that "All alterations take place in conformity with the law of the connection of cause and effect" (B 232), is an explanation of why the category of causality applies to experience. This is essentially part of his meta-metaphysical program. These three tasks, once completed and articulated, will permit the formulation of the set of criteria of adequacy in question. For their completion lays bare the extent to which a priori reasoning plays a role in our procurement of knowledge.

Metaphysical reasoning will consist of a priori reasoning within the confines of experience. But can we be more specific than this? I think that we can, although Kant offers little help. Metaphysics will, in the first place, have a decidedly negative aspect. It will consist of the denial that certain assertions are a part of metaphysics in virtue of either not being synthetic a priori or not ranging over the realm of possible experience. Most of Kant's criticisms of traditional metaphysics occurring in the Transcendental Dialectic of the first *Critique* are metaphysical in this negative way. Metaphysics will, secondly, consist of certain assertions which rest on the categories of the understanding. This suggests that metaphysics' positive task is tracing out the implications of the categories in their application to experience.[4]

In claiming that Kant has meta-metaphysical and metaphysical concerns, I am not, of course, claiming that he always keeps these distinct. Nevertheless, these two features of the theoretical part of the Critical Philosophy are distinguishable and ought to be before us, even if they are not always before Kant. This is especially advisable if we desire to sort out what is significant about the meta-meta-physics/metaphysics distinction for the problem presented by purposiveness in nature.

But the meta-metaphysical and metaphysical endeavors are made all the more intractable by the existence of what Kant calls "ideas," a priori conceptions of a different sort, which also have a use in relation to experience (A 643/B 671). An idea is notion to which no experience can ever be adequate (A 621/B 649). Free Will and God, for example, are ideas. This definition of an "idea" is not, of course, sufficient. For as Kant frequently insists, the same is true of the categories of the understanding. The assertion that "all alterations take place in conformity with the law of the connection of cause and effect" (B 232), the content of the category of causality, "is likewise no concept of experience, precisely because it carries with it the concept of necessity and so of a priori knowledge" (Ak: 455; P: 123).

Ingoring for now that which distinguishes an idea of reason from a category of the understanding, but assuming that they do differ, it is easy to see that ideas call for the meta-metaphysical and metaphysical tasks called for by the categories. The meta-metaphysical enterprise again involves stating what these ideas are, explaining why they apply to experience, and showing how they apply to experience.[5] The metaphysical task again has negative and positive features: denying and affirming that assertions involving these ideas range over the realm of experience. Thus, the existence of ideas not only forces meta-metaphysics to handle ideas in the same way as it handles the categories, it gives meta-metaphysics the additional task of distinguishing these two different sorts of a priori conceptions.

Kant, unfortunately, is not always resolute in this latter task. (1) He makes it clear that the a priori concepts of the understanding, because they are conditions for an object to be known, must apply to experience. As conditions for the "possibility of knowledge" − an expression I will discuss in a moment − they are at the same time conditions for the possibility of that knoweldge called "science"; knowledge of nature, be it science or "common sense," requires these twelve a priori conceptions. In his words, the categories of the understanding are "confirmed by experience and must inevitably be presupposed if experience − that is, coherent knowledge of sensible objects in accordance with universal laws − is to be possible" (Ak: 455; P: 123). (2) He also makes it clear that whereas the categories order the data of the senses, the ideas of reason order the knowledge for which the categories are requisite. "Just as the understanding unifies the manifold in the object by means of concepts, so reason unifies the manifold of concepts by means of ideas..." (A 644/B 672). In fact, the ideas of reason have an "indispensably necessary" employment (A644/B 672). This kind of unifying operation is essential for a body of knowledge to be not just an aggregate, but a system. This goes for both common sense and scientific knowledge. But (1) and (2), as stated, do not tell us much; nor is it clear how they differ. I will try to remedy this situation in the remainder of this section and the next. Later in this chapter, I will indicate how Kant's meta-metaphysical and metaphysical concerns relate to his two conceptions of purposiveness.

In describing Kant's meta-metaphysical enterprise, I have, using Kant's terminology, claimed that he is seeking the "possibility" of certain kinds of judgments. To put the central question of Kant's inquiry into the typically Kantian form, Kant's goal is to answer the question "How are teleological judgments possible?"[6] As is well known, Kant asks a question of the same form in the first *Critique* and *Prolegomena* — "How is knowledge possible?" — to introduce his inquiry there. In his moral writings, his central line of inquiry is drawn by the question "How are moral judgments possible?"[7] The initial source of the obscurity of the Critical Philosophy as a whole can undoubtedly be traced to this methodological procedure. Nevertheless, there are some hints Kant gives which enable us to unpack this loaded question.

The first hint, given by the passage cited from the *Groundwork* earlier, is that in asking about the "possibility" of any kind of judgment (and it is always the possibility of a judgment, statement, assertion, etc. that is in question, though Kant sometimes carelessly talks about objects or events as being possible), Kant is seeking what is *presupposed* by such judgments (Ak: 455; P: 123). The problem is, of course, that the term "presupposition" itself is multiply ambiguous. And on those few occasions when Kant does use the term [*Voraussetzung*] or one of its derivatives, he does little to help clarify what he means in asking how a certain kind of judgment is possible. Thus, we need to hear more than that Kant is seeking the presuppositions of our various judgments.[8]

It might be thought that Kant's own remarks about his method will allow us to close in on the answer we are seeking. Kant suggests that there are two methods by which we may show what is presupposed by our knowledge claims, moral judgments, and teleological assertions: the analytic and the synthetic method. As he describes them in the *Groundwork*, to follow the former, is to proceed "from common knowledge to the formulation of its supreme principle...," whereas to follow the latter, we proceed "from an examination of this principle and its origins to the common knowledge in which we find its application" (Ak: 392; P: 60). As Kant puts it in the *Prolegomena*, in following the analytic method we

> must therefore rest upon something already known as trustworthy, from which we can set out with confidence and ascend to sources as yet unknown, the discovery of which will not only explain to us what we knew but exhibit a sphere of many cognitions which all spring from the same sources (Ak: 4, 274-5; Bk: 23).

Pursuing the synthetic method, we make "inquiries into pure reason itself and endeavor in this source to determine the elements as well as the laws of its pure use according to principles" (Ak: 4, 274; Bk: 21). We can then presumably follow either of these methods to discover how our teleological judgments are possible, i.e., what these judgments presuppose.

Though Kant does not claim to follow either method in the Critique of Teleological Judgment, and in fact combines both there in a most confusing way, these remarks do partially answer our question. For these passages make it clear that it is a *principle* — a synthetic a priori proposition that is the content of an a priori conception — which is sought or shown to be operative by both methods. This suggests that it is a principle that makes teleological judgments possible. This is borne out by noticing that it is the principle of the Second Analogy which in part makes our knowledge claims possible; it is the principle of the categorical imperative that makes our moral judgments possible; it is the a priori principle of taste that makes our aesthetic judgments possible; and it is the teleological principle of the reflective capacity of judgment that makes our teleological judgments possible. Consequently, in seeking the "possibility" of teleological judgments, Kant is searching for the synthetic a priori notions presupposed by such judgments. But what is the nature of this presupposition relation itself? *In what sense* do our knowledge claims, moral judgments, aesthetic judgments and teleological estimations "presuppose" a priori principles?

I believe that the general answer to this question is clear enough. These a priori principles are expressed by or instantiated in our knowledge claims, moral judgments, aesthetic judgments, and teleological estimations; for example, our knowledge claims instantiate the principle of the Second Analogy. Moreover, Kant is intent upon offering some sort of *defense* of the a priori principles instantiated by these four kinds of judgments. The major portion of Kant's Critical program then will consist of isolating the principles expressed in our judgments and then defending them (his metaphysical and transcendental deductions respectively).[9] In each case, once these two tasks are accomplished, the judgments in question become possible. One of the most confusing things about the Critical program, however, is that the *quid juris* question appears to be and perhaps is different in each of the three *Critiques*; the way in which Kant defends the *"legitimacy"* of our four kinds of judgments (given that they each express an a priori principle) varies with each kind of judgment in virtue of the *sort* of a priori principle they instantiate (Ak: 31, 280; B: 122). In the present study I will, of course, confine myself to the nature of Kant's justification of the teleological judgment. In the section which immediately follows I will discuss in more detail the background in terms of which Kant will *defend* the use of the teleological maxim in the context of organic systems. My sketch will quickly lead us from the Critical Philosophy as a whole to the *Critique of Judgment* itself.

2. The quest for unity and the *Critique of Judgment*

One of the more important and recurring themes of the Critical Philosophy is that the human subject and any other subject like it is on a continual quest for unity. From the very beginning of the first *Critique* to the final pages of the third, Kant postulates that the human subject has an innate intellectual need to view whatever he or she is aware of as a part of an inter-connected whole. Kant's assumption that we have an innate interest in viewing things as inter-connected no doubt reveals the influence of Leibniz. In "New System of the Nature and Communication of Substances...," for example, Leibniz postulates "substantial forms" or "primitive forces" on the grounds that "it is impossible to find *the principles of a true unity* in matter alone."[10] That is to say, one of the motives propelling Leibniz' metaphysical program is what he regards as a need to explain the "true unity in matter." It is, of course, not insignificant that whereas in Leibniz the need for unity is dealt with by postulating an elaborate metaphysical scheme, Kant's solution is given in terms of categories and ideas. In moving from Leibniz to Kant, there is a shift from a theory intended to establish metaphysical unity to a theory intended to substantiate an epistemological unity.

Students of Kant's Transcendental Analytic will attest that an intellectual need for unity is assumed throughout his defense of the use of the categories. It is also an interest in seeing things as being connected that is the basis of the theoretical use of ideas. But both sorts of a priori notions bring forth two different sorts of intellectual unity. As we shall see, it is Kant's view that the estimation that purposiveness is present instantiates an *idea* — the idea of design — and the use of this idea will be defended in terms of the cognitive unity it produces. In this section I want to discuss quite generally these two different sorts of unity which the categories and ideas bring forth.

The contrast between the use of an idea and a category is made prominent by Kant's distinction between the determinant and reflective capacity of judgment in the *Critique of Judgment*. In the second Introduction of the third *Critique* Kant makes this contrast in the following way:

> The capacity of judgment in general is the faculty of thinking the particular as contained under the universal. If the universal (the rule, the principle, the law) be given, then the judgment which subsumes the particular under it is *determinant* (even if it, as transcendetal judgment, furnishes a priori the conditions in conformity with which subsumption under that universal is alone possible). But if only the particular is given, where the universal should be found, the judgment is merely *reflective* (Ak: IV, 179; B: 15).

The determinant power of judgment then, "only subsumes under universal transcendental laws which are given by the understanding..." (Ak: IV, 179; B: 15). And these "universal laws without which nature in general (as an object of sense) cannot be thought," the "grounds of the possibility of experience," Kant says, "rest upon the categories" (Ak: V, 182-3; B: 19). The categories are those "modes of knowledge which must have their origin a priori, and which perhaps serve only to give coherence to our sense representations..." (A 2). Thus, the determinant capacity of judgment is the judgmental activity involved in the very production of experience. The determinant power of judgment "constructs" empirical particulars by means of the content of experience and the categories of the understanding. The determinant capacity of judgment then has this feature: it saturates the *content of experience* with the categories. The result is an aggregate of empirical particulars each having certain basic features, such as being subject to the principle of causality. The content of experience becomes, to use a word which will be important for what follows, "systematized" into an aggregate of empirical particulars which are subject to certain basic "principles." Kant's defense of the categories then proceeds in terms of this need to unify the content of experience into a coherent whole.

The theory of the reflective capacity of judgment is Kant's attempt to spell out a second form of cognitive unity.[11] According to Kant,

> *Reflecting* (deliberating) is to compare and hold together given representations either with each other or with the subject's cognitive faculties in reference to a concept which is thereby made possible (Ak: V, 211; H: 16).

The term "representation," which in Kant sometimes means the "object represented" and other times the "representing of the object," here may have either meaning. Thus, the reflective capacity of judgment establishes relations among empirical concepts or conceptualized objects not expressed in the categories, or a relationship between the cognitive faculties which is prompted by an object. The latter activity of the reflective power of judgment, the activity in which "The cognitive powers that are set in play by means of this representation, are found in a free play..." (Ak: 9, 217; B: 52), is the aesthetic capacity of judgment. This essentially *non-cognitive* activity of the reflective capacity of judgment produces a subjective formal purposiveness in nature.[12] The former occupies itself with conceptualized objects and is the *cognitive* activity of the reflective capacity of judgment.

The reflective capacity of judgment when functioning *cognitively* injects a systematic unity into nature by soaking experience *qua* finished product with certain additional concepts, viz., ideas; it functions by providing a second level of conceptualization to that which is experienced. It is the reflective capacity of judgment which is at work when

> the universal is admitted as *problematic* only, and is a mere idea, the particular is certain, but the universality of the rule of which it is a consequence is still a problem. Several particular instances, which are one and all certain, are scrutinised in view of the rule, to see whether they follow from it. If it then appears that all particular instances which can be cited follow from the rule, we argue to its universality, and from this again to all particular instances, even to those which are not themselves given (A 646-7/B 674-5).

Here the

> attitude to this body of knowledge is that it prescribes and seeks to achieve its systematization, that is, to exhibit the connection of its parts in conformity with a single principle (A 645/B 674).

The principle in the case of the reflective capacity of judgment is a description which includes an idea. Thus, the reflective power of judgment (in its *cognitive* mode) performs the additional cognitive function of organizing "the particular laws that can only be made known to us through experience" (Ak: 70, 386; B: 233), by bringing "where possible into the aggregate of empirical laws, as laws, a connection, as in a system..." (Ak: II, 205; H: 11). Just as the determinant capacity of judgment soaks the content of experience with certain basic concepts, i.e., the categories of the understanding, the reflective capacity of judgment when functioning cognitively provides experience with a further layer of intellectual organization. The reflective capacity of judgment functions cognitively by subsuming objects of knowledge under ideas which results in a deeper respect in which known objects may be unified.

One might be somewhat suspicious of describing the determinant capacity of judgment as "saturating" the content of experience and the reflective capacity of judgment as "soaking" experience *qua* finished product. These are, of course, metaphorical characterizations which attempt to underscore the difference between the *determining* character of a *category* and the *non-determining* character of an *idea*. What is the determining character of a category then? To answer this question completely would lead us into the whole of the Transcendental Analytic of the *Critique of Pure Reason*. However, a partial answer will fortunately suffice for our purpose.

The central theme of the Transcendental Analytic is that the concept of objectivity for a creature with a discursive understanding (i.e., a creature which can know only phenomena, not things as they are in themselves) entails descriptions in which a category occurs. (Such descriptions which are pure, i.e., without empirical content, are known as "principles.") For example, consider one of the most important categories: causality. In the Second Analogy Kant argues from the premises that there is objectivity, that natural necessity is a condition for objectivity, and with the causal principle comes natural necessity (and vice versa), it follows that "All alterations take place in conformity with the law of the connection of cause and effect" (B 232). The "principle" of the Second Analogy (one of these "pure" descriptions in which a category occurs) must be included in any Critical account of objectivity. The lesson of the Second Analogy is that part of what Kant will mean by "objective" is "being describable in terms of causal laws."[13] Thus, to say that a description which includes a category is *determining* is to say that the description includes an a priori concept which is an element in the Critical theory of objectivity.

But not only can we describe reality in a determining fashion, according to Kant, we can also describe reality in a non-determining fashion. This occurs when

> Several particular instances which are one and all certain, are scrutinized in view of the rule, to see whether they follow from it. If it appears that all particular instances which can be cited follow from a rule, we argue to its universality, and from this again to all particular instances, even to those which are not themselves given (A 646-7/B 674-5).

The rule to which Kant is referring here is a description in which an idea occurs; it is a description from which we can derive certain true descriptions of reality. Such "rules" thereby organize "particular instances" into a system. However, this "good, proper, and therefore *immanent* use" of ideas is non-determining (A 643/B 671). Indeed, Kant often calls ideas "subjective." Yet this is not to say that descriptions which make use of an idea are illusiory, imaginary, or incorrect.[14] The contrast between categories and ideas is not a contrast between those a priori representations which are objective and those which are subjective. It is rather a contrast between those a priori representations which can be derived from the Kantian theory of objectivity and those which cannot. In calling ideas "subjective," Kant means that none of them, taken singly or jointly, are established in virtue of what "objectivity" means in the Critical system. In other words, an idea can never be shown to be the basis of a principle of the understanding. We can, however, establish the idea, *if* we can "derive the object of experience from the supposed object of the idea" (A 671/B 699). Thus, whereas the concept of an experienced object entails the categories, ideas entail descriptions of experienced objects.[15]

There are two types of *cognitive* activities of the reflective capacity of judgment, the first of which introduces what Paul Guyer calls a "taxonomic" systematic unity into nature,[16] and the second of which introduces what I shall call an "explanatory" systematic unity into nature.[17] In this section we will consider the former. The latter − the explanatory systematic unity of nature − is produced when matter is estimated to be living, and will be discussed in Chapters IV and V.[18]

Kant attempts to describe the taxonomic systematic unity of nature by claiming that it is achieved by following the "*maxims* of reason" (A 666/B 694). These maxims make more precise the conception of nature as a taxonomic system, and are referred to in the Appendix to the Dialectic, where they get their fullest exposition, as "the principles of *homogeneity, specification,* and *continuity* of forms" (A 658/B 686). The maxim of homogeneity, which "keeps us from resting satisfied with an excessive number of different genera..." (A 660/B 688), was followed "when chemists succeeded in reducing all salts to two main genera, acids and alkalies..." (A 652/B680). Another example of the maxim of homogeneity at work was the creation of the Linnaean system of classifying plants and animals. The follower of the maxim of specification "in turn, imposes a check upon the tendency towards unity, and insists that before we proceed to apply a universal concept to individuals we distinguish sub-species within it" (A 660/B 688). In following this maxim, production (*Erzeugung*) is separated into such processes as crystallization, combustion, and generation. The third maxim of reason "combines these two laws by prescribing that even amidst the utmost manifoldness we observe homogeneity in the gradual transition from one species or another, and thus recognize a relationship of the different branches, as all springing from the same stem" (A 660/B 688). This classificatory maxim is at work when the biologist, in studying living organisms, hypothesizes

> an actual relationship between them in the production from a common parent through the gradual advance of an animal genus to another, from those in which the principle of purposes appears to be best articulated, namely man, down to polyp, and from these even down to mosses and lichens, and finally to the lowest step noticeable by us, raw matter (Ak: 80, 418-9; B: 267-8).

These maxims are the guidelines of the knowing subject, be he biologist or brick-layer, for viewing nature as falling into a *classificatory* system. In this way the a priori tendency to view nature as taxonomically systematic can be made more determinate.

But there is, as already indicated, a second *type of cognitive* systematic unity of the reflective capacity of judgment which Kant develops in formulating his explanatory anti-reductionism. It is this sort of systematic unity which is produced when one follows the teleological maxim, the principle instantiated by the estimation that internal purposiveness is present in nature.[19] The systematic unity this maxim produces will be discussed in Chapters IV and V. At least now, however, it is safe to articulate the view the present study will develop of the Kantian project in the Critique of Teleological Judgment in terms of the conceptual features of the Critical Philosophy as a whole.

On the basis of what has already been said, it is clear that Kant's meta-metaphysical investigations conclude that certain a priori concepts have an application to the world which we may experience, either directly, as in the case of the categories, or indirectly, as in the case of ideas. The former, he says, have this application in virtue of being "conditions" for this experience to yield knowledge; the latter are conditions for this knowledge to be "systematic." Both sorts of a priori notions then are conditions for an object to be *known system-*

atically from the point of view of common sense; both are likewise conditions for an object to be known and systematically assimilated by a science. But from the fact that categories and ideas are different sorts of a priori concepts it does not follow that every science relies on the identical set of a priori concepts. This situation will arise, of course, only if not all of the categories or not all of the ideas are necessary for the "construction" of systematic knowledge. Now there is very little reason to believe that Kant thinks that only some and not all of the categories are necessary for all empirical knowledge. (His not believing this, will not affect what follows.) But he does believe that one science might make use of an *idea* which is *not* used by another science. Although the twelve categories of the understanding are necessary for any body of systematic scientific (and common sense) knowledge, all sciences need not share the same set of ideas.

The meta-metaphysical circumstance just described arises with a certain kind of assertion about nature, namely, the assertion that *internal* purposiveness is present. As we shall see, the claim that something is a living thing, an assertion which we might say manifests the *biological point of view*, becomes "possible" by means of an a priori notion missing from the physicist's framework, but present in the biological and common sense conception of the world. This a priori notion is an *idea* and it introduces an *explanatory* systematic unity into nature which I have promised to discuss later. It is accordingly not found within the conceptual framework of physics, biology, or common sense as a condition for knowledge, though the latter two and *not the former* rely on it as a condition for knowledge to be *systematic*. As we would expect, the idea in question is the idea of design. Biology and common sense, insofar as both involve judging that matter is living, become possible through the idea of a design: "the very thought of them [living things] as organized beings is impossible without combining therewith the thought of designed production" (Ak: 75, 398; B: 246). This idea is necessary for the *systematization* of the facts about living phenomena, but not for the systematization of facts about inorganic phenomena. The latter is the kernel out of which Kant's biological anti-reductionism develops.

In the following chapter we will see why Kant would think it a mistake to maintain that intelligent design is a part of the etiology of those natural objects which appear to be the result of conscious design. More specifically, we will see that teleological judgments offer "no ground to presume a priori," as he says, "a special kind of causality" to account for the origin of living systems (Ak: 61, 359; B: 205). In Chapter IV, I argue that Kant is to be viewed as holding that our teleological judgments instantiate a concept which is not present within the mechanical framework of the world. There is, in other words, an a priori principle which these judgments express that is missing from the mechanical conception of the world. This has the consequence, so Kant thinks, that such judgments cannot in some significant sense be "reduced" to purely physical descriptions of nature. Finally, in Chapter V, I pursue Kant's anti-reductionism further by discussing the principle presupposed by the teleological judgment which is absent from the mechanical point of view, and consider Kant's argument that it is absent. In the remainder of this chapter, however, I want to delve a little deeper into the third *Critique* with an eye toward uncovering some of the central themes and goals commonly shared by the Critique of Aesthetic Judgment and Critique of Teleological Judgment.

3. The Critique of Aesthetic Judgment and
   the Critique of Teleological Judgment

The *Critique of Judgment* consists of two parts, the Critique of Aesthetic Judgment and the Critique of Teleological Judgment. In the one the object of Kant's interest is works of art and natural beauty, in the other, the phenomenon of life. What is the common interest which binds two such heterogeneous topics together in one book? Does Kant see something significant in common between works of art and living systems? Do aesthetic descriptions and teleological descriptions share some important features? Is the cognitive process involved in grasping the nature of living matter the same as that involved in grasping the nature of the beautiful and the sublime? I think that these are some of the most intriguing questions one can ask about the Critical Philosophy, and pursuing an answer to them can teach us something about Kant's concern with biological phenomena as well. Consequently, it is to these questions that I wish to devote the remainder of the present chapter.

We may begin by briefly considering the aesthetic judgment. This is not, of course, the place to fully unpack Kant's theory of the aesthetic judgment.[20] Still we need to say something general about

Kant's theory of the aesthetic judgment if we are to achieve the goal we have set out for ourselves in this section. Speaking in terms of aesthetic theory, Kant is an "aesthetic attitude theorist," which is to say that his analysis of the aesthetic judgment entails that in making such a judgment the subject is in a "special" state of mind, often called the "aesthetic attitude." Characterized negatively, the aesthetic attitude is not the attitude or set adopted when one's concerns are practical, i.e., moral or prudential. In Kant this implies, along with certain other key assumptions, that to judge something aesthetically − judge it to be beautiful − the subject has no "*interest*" in the object. This does not mean that the subject is "*uninterested*." The subject is rather closer to the position of the judge who has no "interest" in the case. If he did, he could not be a "*disinterested*" judge, and would be disqualified. But he would not be disqualified if he found the case very interesting. The aesthetic judge is in a position akin to the legal judge. He is not *uninterested*, but *disinterested.*[21] Thus, "disinterestedness" is one of the moments or aspects of the aesthetic judgment.

As we might expect, there are going to be a host of details which accompany Kant's theory (and much controversy over how to formulate these details). However, the key element of Kant's analysis of the aesthetic judgment, I believe, is the assertion that the beautiful pleases without a concept.[22] This statement is, of course, ambiguous. For it might mean that in judging something to be beautiful, I do not possess a concept of the thing or it might mean that I do not make use of my concept of the thing. Since it is clear that I can judge something to be beautiful of which I *do* possess a concept, Kant must mean that in judging something to be beautiful, I do not make use of my concept of the thing. This assertion, which Kant, I believe takes to be axiomatic, is a key premise (along with another for which Kant does not argue, viz., that we find the apprehension of the beautiful to be satisfying) used by Kant to make a number of significant inferences. It is on the strength of these premises (though not these alone[23]) that Kant infers that when I judge something to be beautiful

(1) I possess a disinterested satisfaction with the thing.
(2) The satisfaction I find in the thing is not forced, but free.
(3) The state of mind in which I make the aesthetic judgment can be nothing else than cognition in general; there is an excitement of the imagination and the understanding

through the stimulus of a sensation to indeterminate and harmonious activity.

(4) The aesthetic judgment is subjective.

Let us briefly explain each in turn.[24]

Insofar as the judgment that something is beautiful is a conceptless judgment, and we grant the first *Critique's* doctrine that concepts are a requirement for knowledge, I will not be making use of my knowledge of the object under consideration, I am not aware of what *sort* of thing it is. Its etiology and the role it plays in the practical concerns of my life are of no concern to me, yet I find the object satisfying. Thus, (1) follows, though there remains plenty of controversy about exactly what "disinterestedness" means.[25]

Kant's claim that the aesthetic judgment is not forced, but "free," is established in a similar fashion. Since aesthetic judgments do not involve conceptualization, the predication of beauty to something cannot be grounded on the subsumption of the thing in one class or another. For example, if I subsume something under the concept of dog, I am logically committed to ascribe the predicate "is an animal with a four-chambered heart" to it. In this respect my description of the dog would not be "free." Since there is no conceptualization in the aesthetic judgment, my predication of beauty to a thing will not be forced, but free.

Kant's claim that (3) follows from the two premises looks, however, like a very big jump. Kant is here describing the third moment of the aesthetic judgment, viz., the aesthetic "state of mind" exhibits a subjective formal purposiveness.[26] Kant believes that the fact that the subject makes no use of his concept entails that the mental content of the aesthetic judgment "can be nothing else than... *cognition in general*" (Ak: 9, 217; B: 52).

> The cognitive powers, which are set in play by this representation, are here found in free play, because no determining concept limits them to a specific rule of cognition. Hence the state of mind in this representation must be a feeling of free play of the representative powers in a given representation with reference to a cognition in general. Now a representation by which an object is given that is to become a cognition in general requires *imagination* for the gathering together the manifold of intuition, and *understanding* for the unity of the concept unit-

ing the representations. This state of *free play* of the cognitive faculties in a representation by which an object is given must be universally communicable. For cognition, as the determination of the object with which given representations (in whatever subject) are to agree, is the only kind of representation which is valid for everyone (Ak: 9, 217; B: 52).

This "harmonious activity, viz., that which belongs to cognition in general" (Ak: 9, 219; B: 54), possesses "an inner causality (which is purposive)" (Ak: 12, 222; B: 58), and has the power of *"maintaining the subject in the same state"* (Ak: 10, 220; B: 55). These remarks make it clear that in an aesthetic judgment, the subject's cognitive processes stop short of attaining knowledge. The entire process of experience proceeds as usual *except that the perceived does not become conceptualized.* The mind is, so to speak, just idling, with the cognitive powers — the imagination and the understanding — in a free harmonious play. The mind maintains itself in this "harmonious state" by means of the perception (intuition) of the object.

Though (3) is still obscure (e.g., What is it for the imagination and the understanding to be in harmony?), enough has been clarified to explain the derivation of (4). Since the aesthetic judgment is based upon the harmony of the subject's cognitive faculties, it is subjective. To claim that something is beautiful is, in the last analysis, to make a claim about our own state of mind, i.e., something we conceive to be temporal and not spatial. Since the aesthetic judgment does not "unite the predicate of beauty with the concept of the object," and this is just what would be required, if the aesthetic judgment were objective, it is subjective (Ak: 8, 215; B: 49).

But despite their subjectivity, aesthetic judgments have an alleged universal validity: "[I]f we then call the object beautiful, we believe that we speak with a universal voice, and claim the assent of everyone...." (Ak: 8, 216; B: 50). That aesthetic judgments might be universal is signalled by what Kant takes to be the linguistic fact that

It would... be laughable if a man who imagined anything to his own taste thought to justify himself by saying "This object (the house we see, the coat that person wears, the concert we hear, the poem submitted to our judgment) is beautiful *for me*." For he must not call it beautiful if it merely pleases him (Ak: 7, 212; B: 47).

Thus, aesthetic judgments must be "deduced" transcendentally; their supposed universal validity must be proved. Kant does not disappoint us and attempts such a proof. Roughly speaking, Kant argues that it is legitimate to take aesthetic judgments to be universal on the grounds that they are based upon elements of the transcendental apparatus itself: the imagination and understanding. Since these are universal elements of human cognition which do not vary from individual to individual, any judgment based upon them will be universal.

My account should be considered as but a very brief sketch of Kant's theory of the aesthetic judgment.[27] Still, enough has been said to reveal some of the common threads running through the Critique of Aesthetic Judgment and the Critique of Teleological Judgment. The first and perhaps most obvious is that (a) both are manifestations of what I have called the "Kantian quest for unity." Though I have not tried to unpack Kant's claim that the aesthetic judgment is based upon a harmony of the cognitive faculties, it is clear that this accords with the fact that the human subject is continually seeking unity. In judging something to be beautiful the subject is attaining a kind of intellectual unity, i.e., a unity among his cognitive faculties. In the final chapter of this study we shall also see that the estimation that purposiveness is present in nature is likewise the attainment of a kind of intellectual unity.

Another common thread running from the Critique of Aesthetic Judgment through the Critique of Teleological Judgment is that (b) both the aesthetic state of mind and living phenomena possess the epigenetic capacity of being *self-perpetuating.*[28] As already indicated, according to Kant, the aesthetic state of mind has the power of "maintaining the subject in the same state;" it possesses "an inner causality (which is purposive)." Donald Crawford describes this moment of the aesthetic judgment aptly.

> Beautiful objects tend to maintain and hold our interest.... [T]hey maintain our interest by holding our attention when we are experiencing or contemplating them, to the extent that we tend to ignore, disregard, or even forget about other things. Our perceptual field is narrowed or, as in the case of the laughter of an audience at a play or film, incorporated into our experience to modify the object of our attention. The happenings of the everyday world, our practical concerns, are for the most pushed aside. We become absorbed in attending to the beautiful object....[29]

Just as a living thing is capable of maintaining its physical integrity in the face of various changes in its external environment, so the aesthetic state of mind, according to Kant, possesses the capacity of maintaining itself in the face of external distractions.

In addition, (c) both the aesthetic judgment and the teleological judgment are non-determining. Both, being products of the reflective capacity of judgment, express a priori concepts — ideas — which do not enter into the Kantian theory of objectivity. Kant, unfortunately, does not give an explicit statement of what this idea is in the case of the aesthetic judgment, though it appears that the idea in question is the idea of "assent of everyone" (Ak: 32, 281; B: 123). A discussion of the non-determining character of the teleological judgment will be put off until Chapter V. As we shall see, the taxonomic systematic unity produced by the reflective capacity of judgment is non-determining as well, and this will help explain wherein the non-determining character of the teleological judgment lies.

There is, however, something even deeper common to the two topics of the third *Critique*. The fact that Kant includes a discussion of the aesthetic judgment and the teleological judgment in one book suggests that he believes (d) works of art may be likened to living things.[30] To see why Kant would accept (d), we need to first notice that Kant is one of the forefathers of what is frequently referred to as "formalist theory" in criticism and metacriticism.[31] Formalist theorists view works of art as self-contained entities which detach themselves from their creator at "birth," and thus are to be viewed for what they are and what they are not in their own right. Kant's theory of the aesthetic judgment shares some similarities with the formalist perspective. In making such a judgment the subject does not avail himself of his knowledge of the object. It follows then that the object's features — if it makes sense to speak of "features" of objects now — are considered independently of the artist's intent and our own predispositions toward the object. Only those features of the art object which are currently present to the mind are aesthetically relevant. Notice, however, that were we now to speak of *artistic functions* of works of art or parts of works of art,[32] say the function of a key change in a certain piece of music, such functional ascriptions would of necessity be unpacked in a non-intentional fashion.[33] Functional descriptions of work of art, like functional descriptions of living things (and unlike functional descriptions of artifacts which are not art), will be analyzed in terms of either the goal doctrine or the good consequence doctrine.[34] What I am suggesting then is that Kant's combining of a discussion of the aesthetic judgment and the teleological judgment to form one book is in part due to his belief that (i) both works of art and living systems are teleological, and (ii) works of art and, more generally, beautiful objects are more like living systems than artifacts which are not art in that their teleological character will get unpacked in non-intentional terms.

We have thus uncovered a number of things common to the two parts of the third *Critique*; (a)-(d) seem to indicate that Kant's inclusion of the two topics in the same work is far from arbitrary. Perhaps the question to ask then is not what do the aesthetic judgment and the teleological judgment have in common, but how do they differ? The key to providing an answer to this question is again Kant's assertion that the aesthetic judgment is conceptless.

> Now as a purpose in general is that whose *concept* can be regarded as the ground of the possibility of the object itself; so, in order to represent objective purposiveness in a thing, the concept of *what sort of thing it is to be* must come first (Ak: 15, 227; B: 63).

But

> the judgment of taste is an aesthetic judgment, i.e., such as rests on subjective grounds, the determining ground of which cannot be a concept, and consequently cannot be the concept of a determining purpose (Ak: 15, 228; B: 64).

Both the taxonomic and explanatory systematic unity of nature are constructed with conceptualized objects. That is, they presuppose the work of the determinant capacity of judgment; in short, they are the work of both the reflective and the determinant capacity of judgment. But the aesthetic judgment, on the other hand, is the product of the reflective capacity of judgment alone. Since it is conceptless, the determinant capacity of judgment does not enter into its production.[35]

In addition, there is a respect in which the aesthetic judgment is both non-determining and "subjective," while the teleological judgment is non-determining and "objective." Since an aesthetic judgment does not join the concept of beauty with the concept of an object, it is a judgment about our own state of mind. The aesthetic judgment is about something we conceive to be temporal only, and hence is *subjective*. However, the teleological judgment, while non-determining, is "objective" in the sense that it is about something which is both spatial and temporal.

Thus, though (a)-(d) indicate that the aesthetic and living phenomena share important similarities, there is a crucial difference in the structure of aesthetic and teleological judgments. The motto or "explanation" of the third moment of the aesthetic judgment: "*Beauty* is the form of *purposiveness* of an object so far as this is perceived in it *without any representation of a purpose*" could thus, as stated, suffice as the motto of the Critique of Telological Judgment (Ak: 17, 236; B: 73). But it is to be noted that Kant intends the motto of the third moment to go with the motto of the second: "The *beautiful* is that which pleases universally without a concept" (Ak: 9, 219; B: 54).

Notes

1. In the Preface of the *Critique of Judgment* Kant asserts "For if such a system is one day to be completed under the general name of metaphysic (which it is possible to achieve quite completely and which is supremely important for the use of reason in every reference), the soil for the edifice must be explored by a critique as deep down as the foundation of the faculty of principles independent of experience, in order that it may sink no part, for this would inevitably bring about the down fall of the whole" (Ak: Preface, 168; B: 4).
2. The reader should be warned that this description of Hume is in purely Kantian terms; it is a description of Hume through the eyes of Kant. It is, however, doubtful that the Kantian concept of analyticity can be fit on Hume. See Lewis White Beck, "Analytic and Synthetic Judgments before Kant," in *Essays on Kant and Hume* (New Haven: Yale University Press, 1978), pp. 82-5.
3. As is well known, for Kant a synthetic a priori proposition is one which (i) is informative about the world and (ii) being both necessary (not in a "logical" but a "real" sense) and universal, *has a basis independent of our experience of the world*. See the Introduction of the *Critique of Pure Reason* for Kant's discussion of synthetic a priori judgments.
4. See A xxi and A 81-2/B 107-8.
5. In the first *Critique* Kant echoes this saying "We cannot employ an a priori concept with any certainty without having first given a transcendental deduction of it. The ideas of pure reason do not, indeed, admit of the kind of deduction that is possible in the case of the categories. But if they are to have the least objective validity, no matter how indeterminate that validity may be, and are not to be mere empty thought-entities..., a deduction of them must be possible, however greatly (as we admit) it may

differ from that which we have been able to give of the categories. This will complete the critical work of pure reason..." (A 669-70/B 697-8). Also see the *Critique of Judgment*, Ak: Intro., 1982; B: 19.

6. By "teleological judgment" here and in what follows, I am referring to the estimation that *internal* purposiveness is present in nature.

7. In the *Critique of Practical Reason*, Kant puts it this way: "To decide whether or not something is an object of pure practical reason is only to discern the possibility or impossibility of willing the action..." (Ak: 57; BkR: 59).

8. See Gordon Brittan, Jr., *Kant's Theory of Science*, pp. 28-42 and Robert Paul Wolff, *Kant's Theory of Mental Activity* (Cambridge: Harvard University Press, 1963), p. 45 for a discussion of what Kant might mean by "presupposition" in this context.

9. See "Of the Method of Deduction of Judgments of Taste" (# 31) in the *Critique of Judgment* and "The Principles of Any Transcendental Deduction" in the first *Critique*, A 84-7/B 116-9.

10. Leibniz, *Leibniz: Philosophical Writings*, translated and edited by Parkinson and Morris (London: J.M. Dent and Sons, 1973), p. 116.

11. This second order cognitive unity is also discussed in detail in the Appendix to the Transcendental Dialectic of the first *Critique*. In other words, I am claiming that theoretical reason in the first *Critique* is identical with the reflective power of judgment in the *Critique of Judgment*.

12. I will discuss the subjective formal purposiveness of the aesthetic capacity of judgment in the following section.

13. The story is, of course, much more complicated than this. See Section IV of the Appendix of this work for more details.

14. I owe this point to Ralf Meerbote.

15. One might wonder why an idea's use is always non-determining. Kant's position seems to be that an idea is "excluded" by the very conditions for objectivity (A 677/B 705). What Kant means by this is that the idea is a "concept of an absolute completeness" (A 328/B 385), that is, a "concept of the *totality* of the *conditions* for any given condition" (A 322/B 379). And "since it is the *unconditioned* alone which makes possible the totality of conditions, and conversely, the totality of conditions is always unconditioned..." (A 322/B 379), an idea is both the concept of the absolute (unconditioned) and the complete (totality). The Transcendental Analytic precludes such a concept, for according to the doctrine of the Transcendental Analytic, every objective occurrence is conditioned; it has a cause. From this it follows that there can be no first *event*. Thus, given the truth of the principle of the Second Analogy, no objective occurrence can be *unconditioned*; there is no *absolute*. Likewise, since the causal chain as it proceeds back in time has no end, there is no *totality* of events; there is nothing which is complete. The determining description that every event both conditions and is conditioned has the consequence that descriptions in which ideas occur can at best be non-determining.

16. Paul Guyer, *Kant and the Claims of Taste* (Cambridge, Mass.: Harvard University Press, 1979), Chapter II.

17. I argue that there are these two different theoretical uses of an idea in "A Note on the Thesis Argument of the Third Antinomy," *Ratio*, Vol. 23 (1981), pp. 114-23.

18. One might have expected Kant to use the terms "formal" and "material" to demarcate the two cognitive uses of an idea, but he does not, at least not explicitly. In the First Introduction to the *Critique of Judgment*, however, he does talk about the "formal" as opposed to the "real technic of nature" (Ak: VII, 221; H: 25). By "technic" there he means "purposiveness," but by "formal" and "real" he seems to be referring to internal purposiveness and cases of designedness.

19. One might wonder about what role the three transcendental ideas — "the idea of a

simple self-subsisting intelligence" (A 682/B 710), "the concept of the world in general" (A 684/B 712), and "the idea of *God*" (A 685/B 713) — play in Kant's theory of the reflective capacity of judgment. As I will indicate in Chapter V, in the *Critique of Pure Reason* Kant believes that these ideas are used in the construction of an explanatory systematic unity. In the *Critique of Judgment* the causal maxim and the teleological maxim take the place of the three transcendental ideas.

20. The Critique of Aesthetic Judgment has received considerably more attention than the Critique of Teleological Judgment. For some recent helpful discussions of Kant's Critique of Aesthetic Judgment, see Francis X.J. Coleman, *The Harmony of Reason: A Study in Kant's Aesthetics* (Pittsburgh: University of Pittsburgh Press, 1974), Donald Crawford, *Kant's Aesthetic Theory* (Madison, Wisc.: University of Wisconsin Press, 1974), Paul Guyer, *Kant and the Claims of Taste* (Cambridge, Mass: Harvard University Press, 1979), Eva Schaper, *Studies in Kant's Aesthetics* (Edinburgh: Edinburgh University Press, 1979), and Theodore E. Uehling, Jr., *The Notion of Form in Kant's Critique of Aesthetic Judgment* (The Hague: Mouton, 1971).

21. I am grateful to Peter Kivy for helping me put Kant's position in this way.

22. See sections #4 and 5, and the Second Moment of the Analytic of the Beautiful.

23. This is an important qualifier, for the pleasant satisfies without a concept as well. One of the weaknesses of The Analytic of the Beautiful is Kant's failure to adequately distinguish the beautiful from the pleasurable. Kant does tell us that the pleasurable is based on a sensation, whereas the beautiful is based on a feeling. But *why* this is so is left obscure.

24. I do not think that Kant's theory dictates that (1)-(4) follow in any particular order.

25. See Donald Crawford, *Kant's Aesthetic Theory*, pp. 37-54 and Paul Guyer, *Kant and the Claims of Taste*, Chapter V for two different attempts to explicate Kant's concept of disinterestedness.

26. Important passages describing the third moment occur in the Second Moment of the Analytic of the Beautiful.

27. I shall not discuss Kant's theory of the sublime, since Kant's own remarks (see #30) strongly suggest that his theory of the sublime is logically distinct from the aesthetic judgment and the reflective capacity of judgment.

28. See Chapter I, Section 3.

29. Donald Crawford, *Kant's Aesthetic Theory*, p. 51.

30. Cf. Susanne K. Langer, *Feeling and Form* (New York: Charles Scribner's Sons, 1953), pp. 65, 88-9, and 127-8.

31. A classic exposition of formalist theory is W.K. Wimsatt Jr.'s and Monroe C. Bearsley's "The Intentional Fallacy," in *Philosophy Looks at the Arts*, Joseph Margolis, ed. (Philadelphia: Temple University Press, 1978), pp. 293-306. I am not convinced that Kant really is a formalist. My point is only that others have used Kant to construct formalist theories of art, and of criticism.

32. Hume, in "Of the Standard of Taste," in *Of the Standard of Taste and Other Essays*, John W. Lenz, ed. (New York: Bobbs-Merrill Co., Inc., 1965), p. 16 reminds us that "Every work of art has also a certain end or purpose, for which it is calculated; and is to be deemed more or less perfect, as it is more or less fitted to attain this end."

33. This point is developed in my "Artistic Functions and the Intentional Fallacy," *American Philosophical Quarterly*, forthcoming (May, 1984).

34. See Chapter I, Section 3.

35. Eva Schaper, in *Studies in Kant's Aesthetics*, claims that we can say that aesthetic judgments do "involve categories" without "endangering Kant's characteristic theses about aesthetic judgments" (p. 50). She goes on to claim that "what is distinctive about them in contrast to other judgments of sense perception is that the categories play no justificatory role in assessing validity or invalidity" (p. 50-1). She adds "Notoriously, Kant

overstates his case here, to the extent of saying that concept application plays *no* role in aesthetic judgment. But in saying that the judgment 'this is beautiful' does not apply a concept... to experience, what we should stress is merely that there is no rule for the application of a concept in terms of which the truth of 'this is beautiful' can be intersubjectively established. Aesthetic judgments of taste, we might say, do employ concepts but not in the way that judgments of experience do. The concept of beauty and its associates are not descriptive. The peculiarity of the application of concepts in the making of aesthetic judgments is that when they apply they do so because of something in the individual's awareness of the object, his feeling toward it. Only subjects can feel pleasure and displeasure, and judgments appraising something on the basis of felt apprehension thus differ radically from judgments made on the basis of cognizing something about the object sensed" (p. 51). The question Schaper must answer here is if "concept application" occurs, yet plays no epistemic role, what function do the categories perform in the aesthetic judgment? Telling us that "The peculiarity of the application of the concepts in the making of aesthetic judgments is that when they apply they do so because of something in the individual's awareness of the object, his feeling toward it," does not tell us *why* the categories *apply*, but *how* they would apply *if they in fact did apply*. She in other words offers us a *schematism* of the categories for the aesthetic judgment, not a transcendental deduction of the categories, which is required, or would be required if she is right (see note 4 above). Categories would of course be required for the aesthetic judgment, if there were representations to be unified (as she suggests on p. 51). But Kant tells us not that representations are unified. He tells us that there is a harmony of the cognitive faculties. I think that Schaper goes wrong in her belief that aesthetic claims are a matter of judging (*urteilen*) (pp. 37-50); they are not. They are a matter of estimating (*beurteilen*). The distinction is important (see Chapter I, note 9). Granted, there is the passage at B 141 which suggests that all judgments require the categories. But since aesthetic claims are not judgments, there is no need to require concepts, which is, after all, in strict violation of the explanation of the Second Moment.

CHAPTER III

Design in nature

Finally, according to the right maxim of the philosophy of nature, we must refrain from explaining natural order as drawn from the will of a Supreme Being, because this would not be natural philosophy, but a confession that we have come to the end of it.

Immanuel Kant

1. Is purposiveness designedness?

According to Kant, it seems legitimate to explain the occurrence of events which enter into a causal account of ecological relationships and the parts in a living thing by citing the idea of their effects. Though we have seen that for Kant this use of an idea is *non-determining*, there are those who will insist that purposiveness is in fact designedness. Kant's three theoretical works are scattered with suggestive but chaotic remarks about this tendency to think that purposiveness is designedness, and in the present chapter, we will try to reap some of the insights sown by these remarks. Such an examination will be particularly instructive as a prelude to Kant's attempt to isolate the philosophical significance of our teleological judgments.

Before considering Kant's remarks about the tendency to, as he puts it, "supply to nature causes acting *designedly* and consequently place at its basis teleology... as a *constitutive* principle of the *derivation* of its products from their causes ..." (Ak: 61, 360-1; B: 206-7), the tendency itself should be further specified. This tendency, as I said, is driven by two sorts of judgments about natural phenomena: assertions that there are events whose effects serve as parts of ecological relationships and assertions that there are wholes with epigenetic capacities. However, it is the latter which especially influences this tendency during the seventeenth and eighteenth centuries, and often for reasons other than the teleological features of this occurrence.

It should be recalled that during the seventeenth and eighteenth centuries a great number of European thinkers were beginning to seriously entertain the view that the universe is but a collection of material particles responding to forces in accordance with the laws of mechanics. These laws, which Kant variously referred to as "the laws of motion" or "the laws of moving forces" (Ak: 72, 390; B: 237) are, of course, Newton's three axioms of motion. The axioms assert that (1) every body perseveres in its state of rest, or of uniform motion in a line, unless compelled to change that state by forces acting on it, (2) the alteration of motion is always proportional to the force impressed and is directed toward the line in which the force is impressed, and (3) to every action there is always opposed an equal reaction. To achieve a unified view of the world some mechanistic philosophers such as Descartes, La Mettrie, and Holbach attempted to extend this outlook to the living realm.[1] These materialists confidently maintained that an organism is formed from a mass of particles in accordance with physical laws. In this way every occurrence was thought to be brought about through the action of physical forces.

Clearly the effects of mechanical forces of nature are sufficient to account for the presence of relative purposes in nature. After all, the events which create ecological relationships are the result of a series of causes "working quite undesignedly, causes which rather destroy than favor production, order, and purposes" (Ak: 82, 427; B: 277). As Kant asks

> Now, if the place of habitation of all these creatures, the soil (of the land) or the bosom (of the sea), indicates nothing but a quite undesigned mechanism of its production, how and with what right can we demand and maintain a different origin for these latter products (Ak: 82, 428; B: 278)?

But a biological organism *per se* is both capable of reproducing organisms of its own kind and generating some of its lost or damaged parts. Thus, if biological phenomena behave according to classical mechanics, a little heat, pressure, or friction should suffice to animate a body or bring forth a new limb.[2] Needless to say, empirical observation does not bear this out. Moreover, living organisms apparently do not behave like inorganic objects. For until death, unlike inorganic matter, they withstand much of the oxydizing and destructive effects of their external environment. Warm-blooded animals, for example, maintain their body temperature at a constant level irrespective of the fluctuations of the environment. How then can living processes be absorbed by Newtonian Physics?

For those infected with the new science, one way to deal with the capacities of living things was to dabble in metaphysics, invoking the notion of a Supreme Being to account for these capacities. This led to what was known as the doctrine of preformation, a view held by such researchers as Malphigi, Swammerdam, and Leeuwenhoek.[3] The doctrine, which was frequently called, curiously enough, "the theory of evolution," removed "every individual from the formative power of nature, in order to make it come immediately from the hand of the creator..." (Ak: 81, 423; B: 272).

> For here the mating is a mere formality where a supreme intelligent cause of the world has concluded to immediately form a fruit and only to leave to its mother its development and nourishment (Ak: 81, 423; B: 272).

According to preformation theory, the germ of the living organism, contained in either the egg or spermatozoon, was "like a scale model with all the parts, pieces and details already in position."[4] Fertilization was merely a process of stimulating the preformed being to "evolve," with one parent contributing the actual germ while the other provided the stimulus which activated it and started it growing. Only then could the embryo develop, expand in all directions uniformly, and acquire a final size. Leibniz put it this way.

> Moreover, as the formation of organic bodies appears explicable in the order of nature only when one assumes a preformation already organic, I have thence inferred that what we call generation of an animal is only transformation and augmentation.[5]

This development of the preformed being could then be viewed on the mechanical model, following the laws of motion.

One can now comprehend the sense of "evolution" in this context. For here the "evolution of an organism" means simply the "unfolding of an organism." Moreover, one can see that "evolution," in this sense of the word, is conceivable, particularly when coupled with the concurrent development of the microscope and the discovery of reproductive cells. For the microscope reveals extremely minute creatures that are likely candidates as the preformed individuals. And with the microscope it becomes clear that life beings with a single cell and develops into a mature organism. Though variations exist, the endless cycles of mature organisms developing from one cell suggest that each living thing begins with a minute version of itself.

These preformed beings were thought to be created in a nonnatural way, that is, by a Supreme Being. But notice that the God invoked by the doctrine of preformation was essentially a God of the gaps, not an intelligent designer.[6] The Supreme Being was not called upon here to explain the teleological features internal and external to living things. God's intervention was needed because there was something occurring in nature which science could not handle. Nevertheless, the doctrine of preformation could be argued for on the basis of the teleological features of living things and be one way of attempting to explain the formative capacities of living phenomena.

However, even apart from the general difficulties which emerge in relying on the notion of a Supreme Creator in theoretical accounts of natural phenomena, the doctrine of preformation violated certain standards of reasonableness, elegance, and simplicity. For

> the production of hybrids could absolutely not be accommodated with the system of preformation; and to the seeds of the male creature, to which they had attributed nothing but the mechanical property of serving as the first means of nourishment for the embryo, they attributed in addition a purposive formative power, which in the case of the product of the two creatures of the same genus they would concede to neither parent (Ak: 81, 423-4; B: 273).

Preformation only moved the problem of generation one step backward. Although the existence of the embryo was explained, the formative capacities of the egg and sperm which allowed for the occurrence of growth were not. The sperm in activating the preformed being, and the egg in nurturing it, for example, exhibited the same epigenetic capacities which the Supreme Being was initially called on to explain. To fall back upon Newtonian Physics meant returning to the laws of movement which were just as inadequate for organizing the capacities of the sperm and egg, as for organizing a germ.

Moreover, as Kant indicates, preformation, on the face of it, cannot be reconciled with hybridization. A hybrid, which shares the characteristics of both parents, is the result of an "unnatural" cross. How can preformation accommodate such a cross? For example, suppose that one took a member of one breeding population, a female timber wolf, and a member of a different breeding population, a male airedale, and crossed them. The offspring would have timber wolf traits and airedale traits. The defender of preformation is forced to admit that the little dog already completely formed and present in the timber wolf has airedale characteristics *prior to* the fertilization by the male airedale. Surely this is absurd.

This sort of problem sent its defenders scrambling, taking refuge behind the belief in pre-established harmony between germs. Adherents to the doctrine were of the view then that all organisms, past, presents and future, are formed at the time of creation and awaiting the moment of activation. It was thus assumed that there is a pre-established harmony existing between the egg and sperm coming together at the time of fertilization. The future pairings are established *before* their actual occurrence.

However, perhaps the most serious difficulty with the doctrine of preformation is that it goes beyond the realm of knowledge. As Kant indicates

> Suppose it admitted that a supreme architect immediately created the forms of nature as they have been from the beginning, or that he predetermined those which, in the course of nature, continually form themselves on the same model. Our knowledge of nature is not thus in the least furthered... (Ak: 78, 410; B: 259).

For this reason alone the doctrine of preformation, insofar as it judges living things with

> an objective principle of nature, in accordance with which,

apart from its mechanism (according to the mere laws of motion), quite a different kind of causality attaches to it, namely that of final causes, under which these laws (of moving forces) stand only as intermediate causes (Ak: 72, 389-90; B: 236-7),

ought to be abandoned. The invocation of a "God of the gaps" by the theory of preformation does not explain enough to warrant the claim that there has been or is a Supreme Being operating in the natural world. Yet as mentioned earlier, the claim that there is intelligent design in nature may be argued for by citing empirical data to support the belief that there has been a designer at work in nature. Remaining then is the question of whether there is factual evidence backing the tendency to think that purposiveness is designedness.

2. The empirical question

Besides the belief that the laws of motion are unable to latch on to certain biological phenomena unless we postulate that there has been or is an intelligent designer at work in nature, one might claim that there is independent empirical evidence for thinking that a living thing possesses designedness. This claim that there is design in the universe is a necessary step in the *physicotheological argument* for the existence of God, or as it is more commonly called, "the argument from design." It is thus not surprising to find most of Kant's remarks about the question of design in the universe and the things in it gathered within his various discussions of the physicotheological proof. Let us look then at how Kant construes this proof.

That Kant himself, like most of his contemporaries, is drawn toward the physicotheological argument is clear from his dramatic remarks about the order found in nature. In the first *Critique*, for instance, we find Kant saying

This world presents to us so immeasurable a stage of variety, order, purposiveness, and beauty... that even with such knowledge as our weak understanding can acquire of it, we are brought face to face with so many marvels immeasurably great, that all speech loses its force, all numbers their power to measure, our thoughts themselves all definiteness, and that our judgment of the whole resolves itself into an amazement which is speechless, and only the more eloquent on that account (A 622/B 650).

As we can see, Kant's two conceptions of purposiveness might be viewed as his attempt to capture in detail the order and fitness of living things. Kant then goes on to give a statement of the physico-theological proof.

> (1) In the world we everywhere find clear signs of an order in accordance with a determinate purpose carried out with great wisdom.... (2) This purposive order is quite alien to the things of the world and only belongs them contingently; that is to say, the diverse things could not of themselves have co-operated, by so great a combination of diverse means, to the fulfillment of determinate final purposes, had they not been chosen and designed for these purposes by an ordering rational principle in conformity with underlying ideas. (3) There exists, therefore, a sublime and wise cause (or more than one).... (4) The unity of this cause may be inferred from the unity of the reciprocal relations existing between the parts of the world, as members of an artfully arranged structure − inferred with certainty in so far as our observation suffices for its verification, and beyond these limits with probability, in accordance with the principles of analogy (A 625-6/B 653-4).

Kant rightly sees this proof as an argument by analogy, the conclusion of which is an inference "from the analogy between certain natural products and what our human art produces... appealing to the similarity of these particular natural products with houses, ships, and watches" (A 626/B 654). Kant's two conceptions of purposiveness indicate that the organs of living things, like the parts of a watch, are arranged in such a way that they "exist for the sake of each other"; the ecological relationships of the biological world, likewise, are as perfectly balanced and delicate as the themes of a musical composition. This would seem to justify, he says, the inference that "there lies at the basis of nature a causality similar to that responsible for artificial products, namely, and understanding and a will..." (A 626/B 654).

Perhaps no other piece of analogical reasoning has received more attention than this.[7] Moreover, most of the responses, following Hume's *Dialogues*, have been decidedly critical of this inference. Yet to this move Kant politely responds by saying that such reasoning "probably could not withstand a searching transcendental criticism"

(A 626/B 654). Why, given the fierce attack led by Hume, does Kant nearly let the inference to design in nature slip by?

We can perhaps understand Kant's remark by observing the form of this first stage of the physicotheological proof. To argue that organisms have a designer on the grounds that they share with artifacts a crucial property (being susceptible of functional characterization) is to put forward an argument of the form

F_1:
(1) a is F and D.
(2) o is F.
(3) Therefore, o is D.[8]

Here one concludes that o is D on the grounds that it shares with a the property of F. (On the basis of intelligent design in nature, one would presumably go on to conclude that a Supreme Being exists.) Quite generally, an argument by analogy *infers* that one thing (or sort of thing) is similar to another thing (or sort of thing) in a certain respect in virtue of similarities already *known* to be shared by the two things (or sorts of things). The premises describe the similarities known to exist between the two things (or sorts of things) and the conclusion claims that the two are similar in yet an additional respect. Arguments by analogy are never, logically speaking, valid. However, they may be reasonable though this is not so for F_1. Its critic is quick to point out that artifacts and organisms differ in certain respects. And the *significant* differences between living things and artifacts do not support an analogical argument whose conclusion ascribes some property to living systems. Kant never tires of telling us that living things have a unique formative capacity while artifacts do not. Thus, the critic would caution us not to hastily conclude that some natural objects are designed simply because they share certain features with artifacts. Stated formally, the critic urges

(1) a is D but not C.
(2) o is C.
(3) Therefore, we cannot conclude that o is D.

This sort of criticism, urged by Hume and others, raises serious questions concerning degrees of difference between different sorts of things. More specifically, in evaluating an analogical argument, a judgment must be made concerning the extent, strength, etc. of points of analogy and disanalogy in the things being compared. This underscores the *probabilistic* nature of the argument; it raises the question of whether the fact that artifacts and organisms share a certain trait makes it more or less *probable* that organisms belong to the class of objects which are designed. F_1, given the drastic dissimilarities between organisms and artifacts, is not very probable, and hence not reasonable. To make F_1 plausible we would have to make the further assumption that the presence of F nearly always assures the presence of D. According to Hume, this assumption cannot be granted. However, Hume's criticisms, even if correct, would only show that the conclusion of F_1 is an improbable one.[9]

This sheds some light on Kant's remark. A transcendental criticism is one which is guided by Kant's meta-metaphysical enterprise. Among other things, such a criticism would indicate that an assertion does not range within the confines of experience. Thus, such a criticism would not totally undermine the first stage of the physico-theological argument. It would only show that experience does not support this claim. No one, for example, has ever seen a humming bird being assembled. However, it is not in principle impossible to verify the existence of an intelligent designer of natural objects; in other words, to claim that living things have or had an intelligent designer does not violate any of the principles of the Transcendental Analytic. It is only doubtful that such a being exists. Since a transcendental criticism could at best show that the inference to the presence of design in nature is improbable, it is to be expected that Kant does not promise to blow away the claim that there are causes acting designedly in nature.

Later, waist-deep in the Dialectic of the Teleological Judgment, we find what is perhaps Kant's final word on the matter. Considering once again whether the unique characteristics of the biological realm are the result of "a cause working according to design, i.e., a Being which is productive in a way analogous to the causality of an intelligence" (Ak: 75, 398; B: 245), Kant remarks that "we do not properly speaking, *observe* the purposes in nature as designed..." (Ak: 75, 399; B: 247). Kant then slightly weakens the stand he takes in the first *Critique* saying that we cannot

> judge objectively, either affirmatively or negatively, concerning the proposition: does a being acting according to design lie at the basis of what we rightly call natural purposes, as the cause

of the world (and consequently as its author) (Ak: 75, 400; B: 248)?

To claim that biological processes possess designedness involves making an indirect inference from the observed origins of other sorts of entities, namely, artifacts. And there are equally good grounds for making this inference as for not making it. However, neither grounds are very compelling.

We can perhaps further understand Kant's hesitation in deciding empirically for or against the presence of design in nature by noticing that arguments by analogy are, as Wesley Salmon suggests, "arguments whose function is to evaluate *causal hypotheses....*"[10] The claim that living things have a designer is a theistic causal hypothesis attempting to account for the presence of purposiveness in nature. In keeping with the conceptions of purposiveness and designedness, as well as the latter function of arguments by analogy, rather than viewing the argument for design in nature as exhibiting F_1, we can, and perhaps ought to view it as exhibiting

F_2: (1) a and o are F.
        (2) a's having D explain why it has F.
        (3) Therefore, o is D.

The functional character of human artifacts is explained by the fact that they have a designer. Since biological organisms too are susceptible of functional characterization, we can conclude that the existence of their funtional parts is explained by the operation of a designer.

But even when the argument from design in nature is viewed as exhibiting F_2, the Humean sort of criticism may still be applied; we cannot conclude that o is D for the disanalogies between o and a are too strong. Indeed, organisms possess certain properties — their epigenetic properties — which we *know* explain the presence of some of their other features. Thus, it is unreasonable to accept (3) of F_2; F_2 is also not even a good probabilistic argument. However, if the argument is viewed as instantiating a weaker form

F_3: (1) a and o are F.
        (2) a's having D explains why it has F.
        (3) Therefore, o's having D is a possible explanation for why o has F.

Hume's criticism has almost no effect. F_3 is plausible even though F_2 is not. For (3) of F_3 asserts only that intelligent design is *one* hypothesis accounting for the functional characteristics of living organisms, a hypothesis we know to be true of some functional objects, viz., artifacts. Kant's hesitation in deciding one way or another over the inference to design in nature is thus a sound one. We cannot *prove* that there were or are no causes acting designedly in nature. The empirical objection to this inference then is not final.

One could, of course, raise methodological objections to the claim that an intelligent designer is a part of the etiology of biological processes. But these would be *just* methodological objections. We will look at Kant's methodological objections to this claim in the remainder of this chapter. However, it is probably best to leave the empirical question of whether there has been or is intelligent design in nature with Kant's answer. It is something over which we cannot judge either affirmatively or negatively.

3. Two methodological objections

So far we have only dealt with the first stage of the physicotheological argument, the stage at which the inference to design is made. As already indicated, the proponent of the argument would go on to infer from design in nature a Supreme Intelligence or God. Kant describes this procedure in the closing moments of the Critique of Teleological Judgment.

> The proof which rests on a natural concept that can only be empirical, and yet is to lead us beyond the bounds of nature regarded as the complex of objects of sense, can be no other than that derived from the *purposes* of nature. The concept of these cannot, it is true, be given a priori but only through experience, but yet it promises such a concept of the original ground of nature as alone, among all those which we can conceive, is suited to the supersensible, viz. that of a highest understanding as cause of the world (Ak: 91, 476; B: 329).

Kant's famous criticism of the physicotheological proof in the *Critique of Pure Reason* is almost universally thought to concern the inference not from purposiveness in nature to the presence of design in nature, but from design in nature to a Supreme Intelligent Being, a God. This is the criticism that the physicotheological proof, failing in its attempt to establish the existence of a Supreme Being, falls "back upon the cosmological proof" (A 629/B 657).

> [And] since the latter is only a disguised ontological proof, it has really achieved its purpose by pure reason alone – although at the start it disclaimed all kinship with pure reason and professed to establish its conclusions on convincing evidence derived from experience (A 629/B 657).

It seems to me, however, that there are two ways in which this criticism can be taken, both of which were intended by Kant. The first and most commonly recognized way of interpreting this criticism is as an objection to the claim that a Supreme Being exists. But it is also possible to view this criticism as a *methodological* objection to the claim that some natural entities present us with evidence for thinking that there are "causes acting designedly" in nature. We can best approach the latter by briefly examining the former. This will in turn open up the discussion for a second methodological objection to this tendency made by Kant throughout the Critique of Teleological Judgment.

In order to understand Kant's objection it is helpful to view the physicotheological proof as arguing either

      (1)   There is purposiveness in nature.
      (2)   Therefore, there is designedness in nature.
      (3)   Therefore, there is a Supreme Being.

or

      (1)   Purposiveness in nature is proof that designedness is present in nature.
      (2)   Therefore, there is a Supreme Being.

Kant's complaint is about the inference to a Supreme Being. He maintains that even if we assume that purposiveness in nature proves "the contingency of the form" of things in the world (A 627/B 655),

> that is to say, the diverse things could not of themselves have

cooperated... to the fulfillment of determinate final purposes, had they not been chosen and designed for these purposes by an ordering rational principle in conformity with underlying ideas (A 625/B 653),

it does not prove the contingency of their matter.[11] Purposiveness in nature does not prove that the matter of natural purposes must come from an omnipotent being acting as its cause. For

To prove the latter we should have to demonstrate that the things in the world would not of themselves be capable of such order and harmony, in accordance with universal laws, if they were not *in their substance* the product of supreme wisdom (A 627/B 655).

We would "require quite other grounds of proof than those which are derived from the analogy with human art" (A 627/B 655). We would need to argue that the existence of the matter of natural purposes *in general* proves that there is a Supreme Being. But this is to resort to the cosmological argument. This leads Kant to say:

The utmost, therefore, that the argument can prove is an *architect* of the world who is always very much hampered by the adaptability of the material in which he works, not a *creator* of the world to whose idea everything is subject (A 627/B 655).

The physicotheological proof fails then in its attempt to prove the existence "of an all-sufficient primordial being" (A 627/B 655).

Thus according to Kant, the inference to a Supreme Being in the physicotheological proof is illegitimate unless it also includes a premise to the effect that the designer in nature is not only architect, but creator. Moreover, if the proponent of the argument were to include the premise that the designer in nature is also a creator, he would be putting forward the cosmological argument. Kant's criticism then puts the defender of the physicotheological argument in a dilemma. If he includes in the argument the premise that the designer in nature is also the creator of nature, he is putting forward the cosmological argument, and not the physicotheological proof. If he excludes the premise, he can at best get no further than the past or present existence of an architect in nature.

There is, however, something deeper to Kant's criticism that "the physicotheological proof of the existence of an original or Supreme Being rests upon the cosmological proof..." (A 630/B 658). It can be understood as a *methodological* criticism of the claim that some natural entities are designed. To see this it will be helpful to look at a brief and, surprisingly enough, an overlooked passage in Hume's *Dialogues*.

The remarks in question, made by Philo, are intended to prove "that there is no ground to suppose a plan of the world to be formed in the divine mind... in the same manner as an architect forms in his head the plan of a house which he intends to execute."[12] The reason for this is that if this supposition is made, one is "obliged to mount higher in order to find the cause of this cause."[13] More specifically, since we are talking about a designer's intention as the cause of the presence and arrangement of the parts in a living thing, we are talking about certain mental activities — intending, designing, thinking, etc. — as being causally operative over the material world. Now Philo does not object to such mental-physical interaction, but only notes that "a mental world or universe of ideas requires a cause as much as does a material world or universe of objects; and, if similar in its arrangement, must require a similar cause."[14] Thus, the existence of mental things, like physical things, has a cause. Likewise, just as the order of living things is grounds for asserting an orderer of living things, the order of mental activities is grounds for asserting there to be an orderer of mental activities. As Philo goes on to note

> Nothing seems more delicate, with regard to its causes, than thought.... As far as we can judge, vegetables and animal bodies are not more delicate in their motions, nor depend upon a greater variety of more curious adjustment of springs and principles.[15]

On the assumption that the activity of design is a species of the activity of thinking, Philo asks Cleanthes

> How, therefore, shall we satisfy ourselves concerning the cause of that Being whom you suppose the Author of Nature, or, according to your system of anthropomorphism, the ideal world into which you trace the material? Have we not the same reason to trace that ideal world into another ideal world or new intelli-

gent principle? But if we stop and go no farther, why go so far – why not stop at the material world? How can we satisfy ourselves without going on *in infinitum*? And, after all, what satisfaction is there in that infinite progression?[16]

Philo tightens the screw even further by adding that we might just as well suppose the present material world "to contain the principle of its order within itself" and "assert it to be God," for "the sooner we arrive at that Divine Being, so much the better."[17] It will not do, he adds, "to say that the different ideas which compose the reason of the Supreme Being fall into order of themselves and by their own nature...."[18] We can just as easily say "that the parts of the material world fall into order of themselves and by their own nature."[19]

Philo then, roughly speaking, is arguing

   (1) Let us suppose that where there is order of a certain sort – the functionality, for instance, of the parts of living things and the indispensible relationships of elements of an ecosystem – there are sufficient grounds for claiming that intelligent design is the cause of this order.

   (2) Whenever one engages in the activity of intelligent design one is at the same time engaging in the mental activity of thinking.

   (3) Therefore, where something is the result of intelligent design the designer engaged and/or engages in the operation of thought.

   (4) The mental activity of thinking is ordered; there are parts of this process which are functional (at least in the sense that they are effects which are a means to an end).

   (5) Therefore, where intelligent design occurs there is also intelligent design in its etiology.

   (6) Therefore, the designer of biological organisms and ecosystems must have had a designer, and this second designer a designer as well, *in infinitum*.

(7) Therefore, we cannot explain in a satisfactory manner the presence of natural order by citing the effects of an intelligent designer.

But Philo's argument suffers from a confusion. In saying that the mental activity of thinking depends upon a great variety and "curious adjustment of springs and principles," Philo is confusing the way we are to conceive of the mental activity of thinking with the activity *per se*. For though it may be necessary to conceive of the mental activity of thinking as ordered, the mental activity itself need not have a structure. Indeed, it is the very hypothesis of the mind-body dualist who assumes there to be an intelligent designer in nature that such mental activities may be structureless. In other words, premise (4) is false.

However, even though Philo's argument is defective, it sheds light on how Kant's criticism can also be viewed as methodological objection to the belief that there is intelligent design in nature. For as Kant says in the Critique of Teleological Judgment

> If, however, we assume the purposive combination in the world to be real and to be by a particular kind of causality, namely that of a cause *working designedly*, we cannot stop at the question: Why have things of the world (organized beings) this or that form? Why are they placed by nature in this or that relation to one another? But once an understanding is thought that must be regarded as the cause of the possibility of such forms as they are actually found in things, it must be also asked on objective grounds: Who could have determined this productive understanding to an operation of this kind? This being is the final purpose in reference to which such things are there (Ak: 84, 434-5; B: 284-5).

If we assume that intelligent design is the cause of the purposive combination in nature, we can, and due to our nature, must ask what brought about this intelligent design. The answer is a being which Kant terms "the final purpose," a being which "is unconditioned" (Ak: 84, 435; B: 285). If we attempt to explain the purposive combinations found in nature by citing "a causality similar to that responsible for artificial products, namely an understanding and will" or "a Being which is productive in a way analogous to the causality of intelligence," we are, so to speak, putting ourselves in a position where the only way to explain the presence of the designer of nature is by citing a First Cause of nature. One must put forward a *cosmological argument*. In citing an intelligent designer as the cause of biological order one is not forced to postulate an infinite regress of designers of nature. Thus, the physicotheological proof rests upon the cosmological argument in a second way in that to explain the presence of biological order by citing the activity of an intelligent designer one is forced to fall back upon a First Cause.

Notice that this does not occur when we attempt to explain the purposive combinations found in nature in terms of matter in motion. For if only physical causes are "assumed as the ground of explanation of its purposiveness... it is only the physical possibility of things... that is under discussion..." (Ak: 84, 434; B: 284). If one attempts to explain the presence of purposiveness in nature in terms of particles of moving matter, one need not meet with a First Cause as the ground of these particles. The movement of any collection of particles may be explained (at least in theory) in terms of another collection of particles. The same thing is true when one attempts to account for the purposive features of living things by the means at the disposal of molecular genetics. For although it is true that the DNA of a cell is responsible for the arrangement of protein in a cell, one explains the activity of a cell's DNA in terms of other complex molecules. The latter are likewise explained in terms of other molecules.

Thus, to explain the purposive relations of nature in terms of an intelligent designer is, unlike the physical explanations of science, to have recourse to a First Cause. Certainly such a being has very little cash value in explaining natural phenomena. Consequently, although Kant thinks that we cannot absolutely refute the belief that biological phenomena are the result of intelligent design on empirical grounds, he does believe that this notion is methodologically unsound as an explanation of the occurrence of biological phenomena. For to fall back on an intelligent designer is, as this chapter's opening passage indicates, to give up on natural science. It is rather to head in the direction of a First Cause.

Kant raises another objection to the tendency to think that purposiveness is designedness which he *explicitly* intends to be methodological. Kant first makes this criticism, briefly, in the last section of the Analytic of Teleological Judgment in the *Critique of Judgment*.

If then we introduce into the context of natural science the concept of God in order to explain the purposiveness in nature, and subsequently use this purposiveness in nature to prove that there is a God, there is no internal consistency in either science [i.e., either in natural science or theology] ; and a delusive circle brings them both into uncertainty, because they have allowed their boundaries to overlap (Ak: 68, 381; B: 228).

He expands this same sort of objection in the Dialectic of Teleological Judgment.

And if, starting from the forms of objects of experience, from below upward (a posteriori), we wish to explain the purposiveness which we believe is met with in experience by appealing to a cause working in accordance with purposes, then our explanation is quite tautological and we are only mocking reason with words (Ak: 78, 410; B: 259).

Kant is here urging that explaining the presence of purposive combinations in nature in this manner is in violation of one of the most elementary criteria of adequacy for explanation. To see this, suppose that someone argues for design in nature in the following sketchy manner.

Proof:　(1)　Purposiveness is present in nature.
　　　　 (2)　Therefore, a designer has been at work in nature.

Further suppose that he wishes to explain why purposiveness is present in nature by citing a designer's intention. If we were to make this explanation conform somewhat loosely to the deductive-nomological view of explanation, then it would be construed as the following argument.

Explanation:　(1)　A designer has been at work in nature.
　　　　　　 (2)　Therefore, purposiveness is present in nature.

Now, there is some evidence for thinking that Kant views at least one kind of explanation as fulfilling the DN requirement that a necessary condition for something to be an explanation is that it be capable of being construed as a deductive argument. For he says

> *To explain is to derive from a principle*, which therefore we must clearly know and of which we can give an account (Ak: 78, 412; B: 261; my emphasis).
> In the same natural thing both principles cannot be connected as fundamental propositions of explanation (*deduction*) of one by the other... (Ak: 78, 411; B: 260; my emphasis).

If Kant thinks of some explanations in this way, what he intends by this second criticism is that someone who puts forward the above proof and explanation would be violating a general requirement for any DN explanation, viz., that "the premises, taken singly or conjointly, do not follow logically from the explicandum."[20] Kant is charging then that to attempt both to prove design in nature on the grounds that there is purposiveness in nature and explain purposiveness in nature by citing design in nature is to make the explanans of the DN argument logically equivalent to its explanandum. For purposiveness, according to the proof, entails designedness, while designedness, according to the explanation, entails purposiveness. The result is that the "explanation is quite tautological." The explanatory premises in the explanation assert no more than what is asserted by the explanandum sentence. It would simply be tautological to try to infer the existence of an intelligent cause from the purposive things in nature and then proceed to use that cause to explain them.

Kant is thus not inclined to accept the claim that there is intelligent design in nature. But as should be clear by now, his reason for this does not turn on the question of empirical evidence. Kant accordingly concludes:

> [T]o that end that physics may keep within its own bounds, it abstracts itself entirely from the question whether natural purposes are *designed* or *undesigned*, for that would be to meddle in an extraneous business, in metaphysics (Ak: 68, 382-3; B: 229).

However,

> we speak quite correctly in teleology, so far as it is referred to physics, of the wisdom, the economy, the forethought, the

beneficence of nature, without either making an intelligent being of it, for that would be preposterous, or even without presuming to place an intelligent being above it as its architect, for that would be presumptuous (Ak: 68, 383; B: 230).

If it were possible (which it is not) that natural science could go no further at its level, then it would be permissible to bring the notion of a designer, and hence a First Cause in our explanatory accounts of nature. But until it does, "it must not transgress its bounds in order to introduce... that, to whose concept no experience can be commensurate, upon which we are only entitled to venture after the completion of natural science" (Ak: 68, 382; B: 229).

It is easy to overlook the profundity of Kant's view on the question of design in nature now that we are under the influence of twentieth century science. But Kant rightly saw that although we cannot answer this question negatively, we nevertheless ought not to bring an intelligent designer into our explanatory accounts of nature. This sharply distinguished the scientific point of view from the religious, while leaving his account of purposiveness compatible with both the possibility of divine intervention and the possibility that purposive combinations spring from nature by chance.

Kant's handling of the design argument has two further consequences both of which will be developed in the remainder of this study. First, Kant's criticisms of traditional teleology cut off the inference that living systems are *irreducible* to inorganic phenomena on the grounds that the production of the former involve intelligent design while the production of the latter do not. The sense of "reduction" used here, as well as additional concepts of reduction found in Kant, will be discussed in and seen to be central to the following chapter. Kant's criticism of the argument from design, in the second place, paves the way for the Critical method of dealing with our belief that some natural objects appear to be the result of conscious design. That is to say, the Critical method, rather than taking these judgments as grounds for inferring an intelligent designer, confines itself to their "analysis," an "analysis" which always results in showing the analysandum to instantiate an a priori principle.[21]

78

Notes

1. Three enlightening works discussing the state of biology contemporaneous with Kant are Elizabeth B. Gasking, *Investigations Into Generation 1651-1828* (Baltimore: The Johns Hopkins Press, 1966); Thomas S. Hall, *Ideas of Life and Matter* (Chicago: The University of Chicago Press, 1969), Vols. I and II; François Jacob, *The Logic of Life*.
2. See François Jacob's, *The Logic of Life*, Chapters I and II for a detailed discussion of the phenomenon of generation and Newtonian Mechanics.
3. For a history of preformation theory see Gasking, *Investigations Into Generation 1651-1828*, Chapters III and IV. Kant discusses preformationism in section # 81 of the Critique of Teleological Judgment.
4. Jacob, *The Logic of Life*, p. 57.
5. G.W. Leibniz, *Theodicy*, E.M. Huggard, trans. (New Haven: Yale University Press, 1952), p. 172.
6. Cf. Antony Flew, *God and Philosophy* (New York: Harcourt, Brace and World, Inc., 1966), p. 60.
7. See Antony Flew, *God and Philosophy*, 58-74; David Hume: *Dialogues Concerning Natural Religion*, Nelson Pike, ed. (New York: Bobbs-Merrill Co., Inc., 1970); Alvin Plantinga, *God and Other Minds* (Ithaca, New York: Cornell University Press, 1967), pp. 95-111; Robert J. Richman, "Plantinga, God, and (yet) Other Minds," *Australasian Journal of Philosophy*, 50 (1972), pp. 40-54; and Wesley C. Salmon, "Religion and Science: A New Look at Hume's Dialogues," *Philosophical Studies*, 33 (1978), pp. 143-176.
8. Cf. Paul R. Thagard, "The Best Explanation: Criteria for Theory Choice," *Journal of Philosophy*, 75 (1978), pp. 76-92.
9. Probably the most puzzling thing about Hume's *Dialogues* is what it is that Hume wishes to conclude from them. I believe that the opening quotation of Chapter I is Hume's conclusion.
10. Wesley C. Salmon, "Religion and Science...," p. 145.
11. When Kant says that nature is contingent he accordingly means that an arrangement of facts is inexplicable in terms of natural phenomena. This belief that there is a contingency in the form of some natural objects will be central to the discussions in Chapters IV and V.
12. David Hume, *Dialogues Concerning Natural Religion*, p. 43.
13. Ibid.
14. Ibid..
15. Ibid., pp. 43-4.
16. Ibid., p. 44
17. Ibid.
18. Ibid.
19. Ibid., pp. 44-5.
20. Ernest Nagel, *The Structure of Science*, p. 34.
21. This is, of course, to put the transcendental method in "analytic terms," i.e., in terms of Kant's *analytic method*, discussed in Chapter II, Section 1.

CHAPTER IV

The mechanism of nature

[T]here are patterns of explanation which are indispensable in biology while they do not occur in the physical sciences. These are teleological explanations which apply to organisms and only to them in the natural world, and they cannot be reformulated in nonteleological form without loss of explanatory content.

Francisco J. Ayala

1. Mechanism vs. vitalism, preformation vs. epigenesis

Certainly one of the most striking features of a living thing is its generative and regenerative capacities. It is this characteristic of living things that most impresses Kant. In the previous chapter we saw that the apparent inability of mechanistic science to grasp (in practice) these and other epigenetic capacities of living things motivated the doctrine of preformation, a doctrine which includes the claim that living things are designed by an intelligent being. As we saw, Kant is not sympathetic to the latter claim at all. However, Kant's doubts about what science can do with such phenomena seem to be nearly as drastic as those of the adherents to preformation. For Kant thinks that he sees something about the generative and regenerative capacities of a living thing which leads him to claim that "its form is not possible according to mere natural laws, i.e., those laws which can be cognized by us through the understanding alone when applied to objects of sense..." (Ak: 64, 370; B: 216). Kant goes so far as to say

It is indeed quite certain that we cannot even become sufficiently knowledgeable of, much less provide an explanation of organized beings and their internal possibility according to mere

> mechanical principles of nature; and we can say boldly that it is alike certain that it is absurd to make any such attempt or to hope that another Newton will arise in the future who will make even the production of a blade of grass understandable by us according to natural laws which no design has ordered; we must absolutely deny this insight to men (Ak: 75,400; B: 248).

This is a starting claim. For Kant has already ruled out explaining the occurrence of generation and regeneration in terms of a designer's intention. Moreover, Kant thinks that statements of empirical fact are contingent. And certainly it is a contingent fact that science is unable to explain a certain sort of phenomenon at a given time.[1] Indeed, the significant advances of molecular biology promise, if they do not already offer, an acceptable explanation of generation and regeneration. Why would it be absurd to question whether the production of living matter is not possible according to natural laws?

Much needs to be done in order to unravel Kant's problematic remark. The best way to begin this task is to view his statement in light of the conclusions drawn in earlier chapters. In the first chapter we saw that biological organisms, because of their epigenetic capacities, manifest designer-like and desinged-like qualities. On the assumption that it is legitimate to view these natural things purposively, Kant can be seen as asking what makes this manner of viewing the world *possible*. This question we saw includes the question: "What synthetic a priori principle is instantiated by this point of view?" Thus, to maintain that "the mechanical principles of nature" or "natural laws" do not allow for the internal possibility of living things should entail that there is an a priori principle absent from the mechanical view of nature which the biological point of view instantiates. In the present chapter I will argue that this is in fact how Kant's anti-mechanistic remarks are to be understood. In the following chapter we will consider Kant's *defense* of this a priori principle.

One of the first things to settle on before we proceed is the way we are to interpret phrases like "mechanism of nature," "natural laws," and "mechanical principles." Two senses of "mechanism" come to mind here. As we saw in Chapter III, a mechanist in biology during Kant's time held that all living processes can be adequately explained in terms of the fundamental ideas of classical mechanics of contact action. Thus, when Kant uses the term "mechanist," he might be referring to the Cartesian biologist who equated life with some complex movement of particles in terms of the laws of classical mechanics.[2] Maupertuis (1698-1759), for example, held that the generation of animal life begins with both parents contributing a particulate secretion, and that the new individual is formed from a mixture between the two. The fetus, on this view, then develops gradually as a result of mechanical actions essentially similar to those in inanimate matter.[3] There is, however, a second, more general sense in which Kant may be using the term "mechanist." In this sense a "mechanist" in biology is not necessarily Newtonian. A "mechanist" in this sense holds that all living processes can be explained in a physicochemical manner, that is, by citing the physical and chemical occurrences which are the cause of these processes. (Some might refer to the second type of mechanist as a "physicalist.") Are Kant's anti-mechanistic remarks an indictment against a particular version of physics, classical mechanics, or the physicochemical view *per se*?

It is tempting to think that Kant is here pointing to a certain inadequacy intrinsic to the science of mechanics which will forever prevent it from adequately explaining the generation and regeneration of living things. But this would probably be a mistake. For there does not seem to be any reason why we cannot in principle explain the processes of biological generation and regeneration in terms of the science of mechanics. This, if true, makes it highly unlikely that Kant's anti-mechanistic remarks signal his discovery of a theoretical difficulty with classical mechanics (particularly since he mentions none). Thus, if there is no difficulty of the kind in question, we will have the authorization we need to take "mechanism" in the second and broader sense.

Consider a living system as a system of particles in motion.[4] We can in principle isolate this system's thermodynamical variables: the number of components in the system (that is, the number of particles of matter which comprise the system), the types of aggregations in which the components occur, and the concentrations of the components in each type of aggregation. Thus, under certain conditions each component of the system will occur in various types of aggregations with certain concentrations. The variables of the system are interdependent upon each other, which is to say that the value of one of the variables at any given time may be determined by the values of the other variables. On the basis of these characteristics, there is a sense in which we might "explain" a change in such a system which at the macroscopic level is termed "regeneration." At some initial time t the system is in a definite state (the variables have specific values). In virtue of some change induced in one or more of the variables at t, the system is in some definite though different state after t. One state of the system then is determined by the previous state. Accordingly, the occurrence of the second state can be explained by deducing the proposition that it occurs from a description of the system's initial state and that which induces the change in one of the thermodynamical variables. Insofar as this can be done, one can explain why the second state, the state which is an instance of regeneration, occurs. Admittedly, all this is hypothetical and perhaps impossible in practice (which is just what the defenders of preformation claim). We may never be able to isolate all of the material components of those systems called "living." Nevertheless, this sort of explanation of a biological event is possible in principle. It is highly unlikely then that Kant is saying that the behavior of a living thing *qua* collection of particles is not explicable in terms of the science of mechanics.

Moreover, one can, I think, *understand* the *regeneration* of a trait in a living system from the mechanical point of view by viewing such an event as a mechanical system returning to a state of *equilibrium.*[5] Consider a simple case of a system returning to a state of equilibrium, say a hot water bottle filled to two-thirds of its capacity with water. Assume that no air is present in the bottle and that the water is completely isolated from other disturbing factors in the environment. If pressure on the outside wall of the bottle is increased by pushing on it with one's hand, pressure on the inside wall is increased. When the hand is released, the hot water bottle returns to a state of equilibrium. Likewise, when regeneration occurs in a living thing, it can plausibly be viewed as a mechanical system returning to a state of equilibrium. For example, the observation that a shark is capable of regenerating a lost or damaged tooth may be understood as a mechanical system's capability of returning to a state of equilibrium.

One can extend this notion of equilibrium, though with less plausibility, to the phenomenon of generation itself. Whereas the regeneration of a trait of a living organism may be viewed as a mechanical system returning to a state of equilibrium, the generation of a living thing may be viewed as a species *qua* system of particles returning to a state of equilibrium. For it has long been recognized that a living thing is capable of reproduction when it reaches physiological maturity. At this point a biological organism is in a position to "replace" itself before it dies. Generation may thus be understood as a hypothetical system of particles, which the Cartesian biologist would call a "species," returning to equilibrium.

Admittedly, Kant would not agree that the application of the primitive concept of equilibrium to a mechanical system allows us to *understand* the generation of a living system or the regeneration of a part of a living system. Yet he could not deny that such vital phenomena are explicable in the present sense in mechanical terms. It seems doubtful then that Kant is lashing out against this particular conception of physics because of its inability in principle to cite the *physical conditions* for the occurrence of these biological phenomena. Rather, Kant is making the broader claim that more than the physicochemical conception is *required* to explain *and* understand the epigenetic capacities of living things. Kant's anti-mechanistic remarks are therefore directed toward the mechanical point of view in the second, broader sense in which one may be a "mechanist" in biology.

Nevertheless, though we have fixed the sense of "mechanical" in this context, Kant's anti-mechanistic remarks may yet have an ontological import. For in claiming that it is impossible to explain living processes in terms of "natural laws," it might be thought that Kant has adopted some version of vitalism. Vitalism, generally speaking, is the view that a living thing is not only made up of physical inanimate parts, but also consists of a non-material entity which brings with it the activities characteristic of living organisms. This vital entity, in animating the organism, distinguishes the organic from the inorganic. Thus, it might be claimed that although Kant feels that living things are in part comprised of physicochemical matter, he believes that they possess something else. This is a vital entity which is neither physical nor chemical in nature.

Certainly if Kant were a vitalist, we would have a quick solution to the problem of interpreting his remarks about the inexplicability of biological organisms. Moreover, the fact that Kant cites with approval the work of Johann Friedrich Blumenbach, a famous eighteenth century physiologist and vitalist, might be thought to provide a reason to believe that Kant's remarks have a vitalist intent.[6] Blumenbach opens his *Institutiones Physiologicae* with the statement:

> In the living human body, the healthy functions of which con-
> stitute the exclusive object of the science of physiology, there
> are three things worthy of our immediate attention and regard;
> namely, The solids, or parts containing; The fluids, or parts
> contained within the solids; And lastly, the Vital Energies,
> which in the consideration of the defense of physiology, con-
> stitute the most interesting and important objects of our regard.
> It is in consequence of these energies that the solids are ren-
> dered alive to the impulse of the fluids, endowed with a power
> to propel the same, and also to perform a variety of other
> motions.[7]

Among the vital energies (which do not possess physical, chemical, or
mechanical properties), the most important is "the nisus formativus
or formative propensity, which should be considered as the efficient
cause of the whole process of generation (taken in so extensive a
latitude as to include both nutrition and re-production [regenera-
tion] as modifications itself)."[8] The necessary function of the nisus
formativus is to cause the epigenetic development of the mingled
sexual fluids into a zygote. It is thus "the source of all generation,
nutrition, and re-production [regeneration], in each organized king-
dom."[9]

In discussing the epigenetic capacities of living things Kant remarks that no one has done more to develop the theory of epigenesis
than Blumenbach. Kant continues:

> That crude matter should have originally formed itself accord-
> ing to mechanical laws, that life should have sprung from the
> nature of what is lifeless, that matter should have been able to
> dispose itself into the right form of a self-maintaining purposive-
> ness· — this he rightly declares to be contradictory to reason.
> But at the same time he leaves to natural mechanism, in its sub-
> ordination to this inscrutable *principle* of a primordial organiza-
> tion [the nisus formativus], an undeterminable but yet unmis-
> takeable share... (Ak: 81, 424; B: 274).

Thus, it might seem that Kant's anti-mechanistic remarks are based
on a doctrine such as Blumenbach's.

In the Critique of Teleological Judgment, however, it is clear that
Kant is not a vitalist. For in postulating the presence of a vital entity

which is "the efficient cause of a the whole process of generation," we run up against the same sort of difficulty which confronts the physicotheological tendency discussed in the last chapter. To fall back on a vital entity to explain the generative, and the rest of what I have termed the "epigenetic" capacities of living things, is to fall back on that which science cannot pursue. To say that a warm-blooded animal retains a constant temperature in spite of temperature fluctuations in its environment because of the presence of a vital energy in the animal is, like postulating an intelligent designer, to take refuge in a realm where science cannot enter. Indeed, it would seem that the only way to account for the presence of these vital forces is by citing a First Cause. Moreover, in discussing these epigenetic capacities of living things Kant is critical of the vitalist position.

> We perhaps approach nearer to this inscrutable property if we describe it as an *analogue of life*, but then we must either endow matter, as mere matter, with a property that contradicts its very being (hylozoism) or associate with it a foreign principle *standing in communion* with it (a soul). But in the latter case we must, if such a product is to be a natural product, either presuppose organized matter as the implement of that soul, which does not make it a whit more comprehensible, or regard the soul as the artist of this structure, and so remove the product from corporeal nature (Ak: 65, 374-5; B: 221).[10]

There is no reason then to think that Kant's anti-mechanistic remarks are tantamount to an acceptance of vitalism.[11] His agreement with the vitalist Blumenbach is a conceptual one only. The latter, in postulating a nisus formativus, acknowledges that living things manifest designer-like and designed-like properties. Kant's praise of Blumenbach is an approval of this acknowledgement.[12]

At this point it will be helpful to draw together the doctrines of preformation, epigenesis, mechanism, and vitalism as they were expounded in the eighteenth century. During this time there were two central debates concerning the nature of living matter which Kant witnessed: the debate between mechanism and vitalism and the debate between preformation and epigenesis. These two conflicts were not unrelated. though their relationship was far from a straightforward one. Let us consider the latter controversy first.

The debate between preformation and epigenesis was a quarrel between those who removed "every individual from the formative power of nature, in order to make it come immediately from the hand of the creator..." (Ak: 81, 423; B: 272), and those who regarded "nature as self-producing, not merely as self-evolving [self-unfolding]" (Ak: 81, 424; B: 273). In other words, preformation is the doctrine that the purposive combinations in nature are constructed *outside of nature* by a Supreme Cause, while epigenesis is the doctrine that these combinations are constructed *within nature* by the living thing itself. Mechanism vs. vitalism, on the other hand, was a debate between those who believed that any biological event occurring in a particular living system is caused by and is thus explicable in terms of some mechanically characterizable occurrence, and those who believed that biological phenomena, such as generation and regeneration, occur as a result of the causal efficacy of a vital entity.

These two different debates become intertwined by the fact that both mechanist and vitalist may hold the doctrine of epigenesis. Thus, both Maupertuis, a mechanist, and Blumenbach, a vitalist, could be in agreement over how a living system should be conceptualized, i.e., as a self-producing system, but disagree about the nature of these systems.[13] Consequently, preformation can be in conflict with either mechanism or vitalism for conceptual reasons.[14] But epigenesis and preformation can also clash on ontological grounds. Such a collision occurs when the defender of epigenesis is a mechanist. For the doctrine of preformation admits that the *First Cause* is the cause of the zygotic substance (which has existed preformed since the beginning of time), whereas the mechanist holds that the cause of the zygotic substance is physical in nature.

Kant's philosophical characterization of living things is developed in the context of these two debates. As we saw in Chapter I, Kant conceives of living systems epigenetically; he believes that "nature has formative powers."[15] Yet, as we have seen, he rejects both preformation theory and vitalism. Nevertheless, there is something which Kant finds deeply troubling about explaining "even the production of a blade of grass" in physicochemical terms. And this gets built into his theory of the teleological judgment.

2. Reductionism in Kant

It is apparent then that Kant is not only in some sense an "anti-mechanist," but he is also an anti-vitalist. He can accordingly be viewed as adopting a position often referred to as the "organismic" standpoint, a position which affirms the *irreducibility* of biology to physics. But if Kant's anti-mechanistic remarks are to be considered as a denial that biology is reducible to physics, what sort of anti-reductionist is he?

On the basis of the ground covered so far, as well as other remarks Kant makes, we can safely say that Kant is not anti-physicalist, i.e., his anti-mechanist position is not ontological. Thus, we may say that Kant is not an *ontological* anti-reductionist. By *ontological reductionism* I mean the view that every objective event is one which we can in theory describe in physicochemical terms, and one which occurs in virtue of being caused by some other event we can in theory describe physicochemically.[16] Thus, ontological reductionism entails that such biological processes as generation and regeneration are describable and have causes describable in physicochemical terms. According to this version of reductionism, it is true to say that living processes are just complex and special patterns of physical and chemical processes. Those who put forward the doctrine of preformation and vitalism are ontological anti-reductionists insofar as they hold that biological processes have a cause (an intelligent designer or vital entity) indescribable in physicochemical terms.

In the Critique of Teleological Judgment, as we have seen, Kant is critical of both of these ontological anti-reductionist positions. In fact, although not admitting outright that he is an ontological reductionist, Kant does indicate that he has no objections to this brand of reductionism.

> For example, it may be that in an animal body many parts can be conceived as concretions according to mere mechanical laws (as hide, bone and hair). And yet the cause which brings together the required matter, modifies it, forms it, and puts it in its appropriate place, must always be estimated teleologically, so that here everything in it must be considered as organized, and everything in it, again in a certain relation to the thing itself is an organ (Ak: 65, 377; B: 224).[17]

This passage indicates that Kant's anti-mechanistic remarks are not a denial that living processes are anything but the sort of thing studied by physics. He only states here that it is necessary for us to *view* a living thing from a point of view that is different from the mechanical conception of nature. (More will be said about this point of view later in this chapter.)

There is a second type of reductionism which is discussed from time to time by Kant. This variety of reductionism, which might be termed *"methodological reductionism,"* is one instance of Kant's well-known heuristic view that we *ought* to examine nature *as if* a certain doctrine about nature is true. More specifically, methodological reductionism concerns the manner of investigation which ought to be employed in the study of living phenomena. The extreme methodological reductionist would contend that one ought always investigate living processes in terms of the mechanism of nature only. That is to say, he would urge that biology ought to proceed only by studying the physicochemical nature and etiology of the processes of generation and regeneration. The other extreme is the methodological anti-reductionist. For example, much has been learned about living things by following the Mendelian approach in the study of hereditary phenomena. Mendelian genetics proceeds by studying the phenotypic traits resulting from various crosses within a species while ignoring the physicochemical nature of genotypic structures. Were we to find a geneticist who held, in the face of the impressive discoveries made by molecular genetics, that the only way hereditary phenomena should be studied is by a Mendelian approach, we would have a methodological anti-reductionist in genetics. But one can admit that living phenomena should be studied mechanically while maintaining that this is not the only way we should investigate vital processes.

That Kant is to some degree a methodological reductionist is clear. For he believes that in the study of vital phenomena

> I *must* always *reflect* upon them *according to the principle* of the mere mechanism of nature, and consequently investigate this as far as I can, because unless this lies at the basis of investigation, there can be no proper knowledge of nature at all (Ak: 70, 387; B: 234).

But Kant is not an extreme methodological reductionist in that he believes that investigations into living processes ought to proceed from a different angle as well. This different perspective is, of course, the teleological.

There is, however, a third sort of reductionism present in Kant's discussions of vital phenomena to which he does strenuously object. Moreover, in the course of this chapter we shall see that this anti-reductionism is the source of Kant's anti-mechanistic remarks. We shall see that his anti-mechanistic stance stems from his belief that there are concepts embedded in our biological conceptual scheme — concepts of design — which cannot be constructed from the conceptual resources of physics. The very thought of living things, Kant stresses, "as organized beings is impossible without combining therewith the thought of their designed production" (Ak: 75, 398; B: 246). This will in turn make the reduction of biology to physics impossible, or to put it in a more Kantian way, it will show that the mechanical point of view is incapable of providing an answer to the question "How are teleological judgments possible?"

Perhaps the best way to begin characterizing Kant's anti-reductionism, a task which will take the remainder of this study to complete, is to determine whether this irreducibility of biology to physics is temporary or permanent. The two possibilities might be developed in the following way.[18] (1) In making these anti-mechanistic remarks Kant is pointing to the incipient stages of the biological sciences during his time. However, as biology develops, this deficiency may be removed (since it would, after all, be hasty to put limits on the advancements of science). Granted, teleological explanations (or as Kant sometimes prefers to put it, making use of "concepts of design") may be necessary until mechanistic ones are available, but there is no reason in principle why the former should not in the end be dispensed with by science. When we do adopt the teleological point of view, however (and we do this when non-teleological laws are inadequate), it is impossible to explain living things in terms of the mechanism of nature. Thus, on this interpretation, Kant's anti-mechanistic remarks just amount to the claim that the teleological point of view (which is here understood to consist of explaining the presence of any natural occurrence in terms of an intelligent designer or vital force) is incompatible with the mechanical. They are incompatible insofar as the mechanistic point of view attempts to account for every natural occurrence physicochemically, whereas the teleological point of view, as it is understood here, falls back upon an entity whose existence can only be explained by citing the causal efficacy of the First Cause.[19] Whenever we adopt the latter, it is absurd to try to explain and cognize living things from the mechanical point of view. (2) There are permanent conceptual features of the biological point of view which make it absurd to attempt to provide a pure mechanical account of organized beings. Kant is pointing to a fundamental and permanent difference between the physical and biological sciences due to conceptual elements unique to the biological point of view.

Consider (1) first. According to (1), Kant is not claiming that there is something permanently irreducible about living things. In fact, biology will be reducible as soon as its physicochemical conception becomes sufficiently refined. The only absurdity involved in attempting such a reduction is trying to do so while viewing the world teleologically. For to view something teleologically here is to view it as the effect of an event incapable of being characterized physicochemically. It is interesting to note that some philosophers and scientists today think that the "reduction" of biology to chemistry and physics in a sense other than "ontological" or "methodological" is not possible at the moment, though possible in principle. Ayala is representative of this view.

> It is clear that in the current state of scientific development a majority of biological concepts such as cell, organ, Mendelian population, species, genetic homeostasis, predator, tropic level, etc. cannot be adeqautely defined in physicochemical terms. Nor are there at present any class or classes of statements belonging to physics and chemistry from which every biological law could be derived.... These considerations make it clear that the reduction of all or most of biology to the physicochemical sciences is premature at present.[20]

Are Kant's anti-mechanistic remarks simply an indication that he thinks that the teleological and mechanistic viewpoints are incompatible? Does he hold, like Ayala, that the reduction of biology to the physical sciences is at least a practical possibility?[21]

It is clear that Kant regards the teleological and mechanical perspective as somehow incompatible. The question is whether Kant is viewing teleological reasoning, as one commentator puts it, "as a hypothetical mode of reasoning required by our inability to give a complete causal account of a certain kind of facts namely facts of organization in living processes."[22] Or is he indicating that the mechanical point of view is forever banned from giving an adequate account of living processes? The evidence weighs heavily in favor of thinking that it is the latter, (2) above, that is intended by Kant's remarks. For it is Kant's belief that

> We are in fact indispensibly obliged to ascribe the concept of design to nature if we wish to investigate it, though only in its organized products, by continuous observation; and this concept is therefore an absolutely necessary maxim for the empirical use of our reason (Ak: 75, 398; B: 245-6).

In fact, Kant asserts

> *According to the constitution of the human understanding*, none other than designedly working causes can be assumed for the possibility of organized beings in nature; and the mere mechanism of nature cannot be adequate to the *explanation* of these products (Ak: 78, 413; B: 262; my emphasis).

This leaves little doubt that Kant thinks that the *irreducibility* of biology to physics, due to the *inexplicability* of living things in terms of physics, is permanent.

But these remarks exhibit an important feature of Kant's anti-reductionism. Since Kant maintains that we cannot explain living organisms in terms of the mechanism of nature, while not objecting to the view that we can in principle cite the physicochemical causal conditions of a biological event and in this sense explain its occurrence, he must believe that there is a second sense in which we can provide an "explanation" of something. Thus, Kant's remarks about the "inexplicability" of organized beings make use of a second concept of explanation. Accordingly, the third sort of reductionism discussed in the Critique of Teleological Judgment might be termed *"explanatory reductionism."*[23] An explanatory anti-reductionist like Kant holds that though we can "explain" generation and regeneration mechanically in the sense that we can cite the physicochemical causal conditions for their occurrence, there is another significant sense in which the mechanical view of nature cannot "explain" these occurrences. In the following chapter we will see that the incompatibility between the teleological and mechanical point of view is rather a conflict between two competing *modes of explanation* in this second sense. That is, biology requires a mode of explanation which cannot get introduced by the mechanical point of view.

Consequently, when Kant claims that the cognition of living things requires concepts of design, he means that there is something missing from the mechanical point of view which does not permit it to "explain" living phenomena in this second sense we have yet to uncover. The following chapter will in part consist of an examination of this second concept of explanation. In the remainder of this chapter, however, we will complete the preliminaries for the next by further exploring Kant's statement of his anti-mechanist position.

3. Kant's anti-reductionism

It is Kant's belief that living things are purposive in character, which is to say that they *form* their parts, in addition to *being caused by* their parts. Moreover, the epigenetic character of living things makes it necessary to view them in terms of the idea of design. We have seen that in asserting that living phenomena must be viewed in terms of this idea Kant is not maintaining that living things are in fact designed. Indeed, the non-determining character of an idea, Chapter III, and the discussion of vitalism in this chapter rule out this possibility. In the following chapter we will consider what is involved in viewing living phenomena in terms of the idea of design and consider Kant's argument for this claim. In the remainder of this section, however, we will consider Kant's *explication* of these concepts of design, and in the following section, we will attempt to understand why Kant believes that the mechanical conception of nature lacks the materials needed for their construction.

Our exploration of Kant's idea of design may begin by returning once again to the early sections of the Critique of Teleological Judgment. Here Kant distinguishes between two different causal relations.

> Causal combination as thought merely by the understanding is a connection constituting an ever progressive series (of causes and effects), and things which as effects presuppose others as causes cannot be reciprocally at the same time causes of these. This sort of causal combination we call that of effective causes... (Ak: 65, 372; B: 219).

[A] causal combination according to a concept of reason (of purposes) can also be thought, which regarded as a series would lead either forward or backward; in this the thing that has been called the effect may with equal propriety be termed the cause of that of which it is the effect. In practical matters (namely the arts), we easily find connections such as this; e.g., a house, no doubt, is the cause of the money received for rent, but conversely the representation of this possible outcome was the cause of building the house. Such a causal connection we call that of final causes... (Ak: 65, 372; B: 219).

Thus, according to Kant, in causal relations of final causes the effect does not actually generate the cause. Rather the *thought* or representation of the effect is the cause. It is in virtue of the *thought* of the effect causally generating the cause that we can with propriety speak of the effect as "the cause of that of which it is an effect." Both final and effective causes are causes which proceed in one direction. The effects in causal combinations of final causation cannot, like the effects in causal combinations of effective causes, *be the cause of the initial cause*. They can only be *regarded* in this way.

It is Kant's belief that "certain things of nature (organized beings) and their possibility must be estimated in accordance with the concept of final causes..." (Ak: 72, 389; B: 236-7). Moreover, it is apparent, as the following passage indicates, that for Kant the concept of final cause is bound up with the concept of design.

Now if it is asked why a thing exists, the answer is either: its presence and production have no reference at all to a cause working according to design, and so we always refer its origin to the mechanism of nature; or: *There is somewhere a designed ground of its presence* (as a contingent natural being). *And this thought one can hardly separate from the concept of an organized being: since we must place at the basis of its internal possibility a causality of final causes* and an idea at the ground of this, we cannot think the existence of this product except as a purpose. For the represented effect, whose representation is also the determining ground of the intelligent working cause of its production, is called a purpose (Ak: 82, 425-6; B: 275; my emphasis).

To say that our view of living processes is in part comprised of concepts of design is to say that we see causal combinations of final causes within these processes. Once it is recalled, to cite a passage quoted earlier, that "We are in fact indispensibly obliged to ascribe the concept of design to nature if we wish to investigate it, though only in its organized products, by continuous observation..." (Ak: 75, 398; B: 245-6), we approach the core of Kant's anti-reductionism. For the difference between the biological point of view and the mechanistic lies in the fact that there is a concept of causality found in the former which is not present in the latter. Biology, it seems, is not reducible to physics because in cognizing living matter one makes use of a concept of causality not present in the mechanistic point of view. The biological point of view makes use of the *idea* of a final cause.

But why is it that Kant thinks that the concept of a final cause is not "reducible" to the concept of an effective cause? Given that the only significant difference between these two sorts of causal combinations is that in final causation there is an "intelligent" effective cause, why should the mechanistic point of view be different from the biological? Consider a case of intentional human action. Picture a woodsman constructing a log cabin by notching out with an ax the ends of each log so that they will support the weight of the cabin. The woodsman's action, swinging the ax, causes the notch to be formed at the end of the log. This is apparently a case of final causation, since the representation of the effect is part of the cause of the woodsman's action. Why is this causal relation of a different *kind* than that which is instantiated by a rock-slide crashing down onto and notching out the ends of some freshly cut logs?

One might be tempted to say that in the former case there is a mental event causing a physical event, whereas it is just two physical events which occur in the latter. Thus, pursuing this line of thought, one might claim that in mechanical processes one finds only physical events, but in biological processes there are mental events, i.e., *immaterial* events which causally interact with physical events. But this possibility is ruled out automatically if Kant is an anti-vitalist. If the epigenetic activity of a living thing is thought to be a mental activity, then Kant would be committed to vitalism. For living things would then be construed as comprised in part of things not physicochemical in nature. Moreover, as we saw in the previous section, Kant has no objections to ontological reductionism, a view which asserts that all stuff in the natural world is physicochemical in nature. If so, the woodsman's thought about notching out the end of a log can be construed as a system of particles in a definite state. There seems to be no reason to think that final causation is anything but effective causation then, and thus no reason to think that the biological point of view is irreducible to the mechanistic.

We can isolate why Kant thinks the idea of a final cause is not reducible to the concept of an effective cause, and thus what Kant thinks is irreducible about living things, however, by exploring his statement that final causes are "possible only through reason," which he says is "the faculty of acting in accordance with purposes (a will)" (Ak: 64, 370; B: 216). Whereas final causation is the causality of reason, we have seen that efficient causation is the "causal combination as thought merely by the understanding." That is to say, final causation is the notion of causality found within Kant's three moral works and effective causation is the notion of causality present in the Transcendental Analytic of the first *Critique*. Now perhaps one of Kant's most infamous doctrines is that there is a sense in which these two types of causality are mutually exclusive within the natural world, i.e., a final cause cannot originate (though its effects can appear) in the natural world. Consequently, if our conceptual scheme for estimating living things involves the concept of final causation, then living things are irreducible to physicochemical phenomena. For the causal relations between the latter sort of things are always effective *only*.[24] Biology will be seen to be autonomous then, once Kant verifies that living phenomena enter into causal relations of final causes while mechanical processes do not. Of course, the notion of a final cause is at this point just as obscure as Kant's concept of design. But we can begin clarifying this notion by briefly glancing at these two concepts of causality as they emerge in Kant's first *Critique* and three moral works.

Let us turn to the Second Analogy of the first *Critique* in order to refresh our memory of the conception of causality Kant is working with in the *Critique of Pure Reason*. In the Second Analogy Kant's primary purpose is not to explicate the concept of causality. It is rather to show that

Experience itself — in other words, empirical knowledge of appearances — is thus possible only insofar as we subject the succession of appearances, and therefore all alteration to the law of causality; and as likewise follows, the appearances as objects of experience, are themselves possible in conformity with the law (B 234).

It is the notion of causality that Kant uses in attempting to solve this problem that is our concern.

The formulation of "the principle of the Second Analogy" found in the first edition of the *Critique of Pure Reason*, the formulation which is better for our purposes, is referred to as the "principle of production" (A 189). This is the rule or principle that "Everything that happens, that is, begins to be, presupposes something upon which it follows according to a rule" (A 189). That Kant here is referring to effective causal relations is clear, for he says

> in an appearance which contains a happening (the preceding state of the perception we may entitle A, and the succeeding B) B can be apprehended only as following upon A; the perception A cannot follow upon B but only precede it (A 192/B 237).

Kant here emphasizes, as he does in the Critique of Teleological Judgment, that in a causal relation the effect can never be temporally prior to its cause. Kant, moreover, maintains that causal succession occurs according to rule, and this rule "makes the order in which the perceptions (in the apprehension of this appearance) follow upon one another a necessary order" (A 193/B 238).[25]

> The situation then, is this: there is an order in our representations in which the present, so far as it has come to be, refers us to some proceding state as a correlate of the event which is given; and though this correlate is, indeed, indeterminate, it nonetheless stands in a determining relation to the event as its consequence, connecting the event in necessary relation with itself in the time series (A 198-9/B 243-4).

Thus, according to Kant's concept of effective causation, a cause is that event from which an occurrence invariably follows. Every event must be seen as the successor to some previous event upon which it follows with necessity; it occurs at this determinate point in time *because* something else occurred previously. Every effect follows from its cause then with necessity, and thus every event in nature is a link in a necessary chain of events. What *kinds* of events are linked in this way is established empirically. Nevertheless, every natural occurrence follows necessarily from its cause.

Kant's concept of effective causality is quite different from the concept of causality found within his moral works. To see this, we need to refresh our memory of the notion of causality operative within Kant's theory of morality.

Kant's analysis of moral action is quite intricate, but we need not pursue it in detail for our purposes. We can quickly isolate the features of the notion of causality involved here by looking at the central kind of moral behavior, what Kant in the *Groundwork* calls acting "*from duty*" (Ak: 397; P: 65). In acting from duty "nothing but the representation *of the law* in itself... is the ground determining the will..." (Ak: 401; P: 69). The result is that

> An action done from duty has to set aside altogether the influence of inclination, and along with inclination every object of the will; so there is nothing left able to determine the will except objectively the *law* and subjectively pure reverence for this practical law... (Ak: 400; P: 68-9).

What makes the action morally right, what determines "its objectivity," is likewise that which summons it into existence, viz., the moral law.[26] Kant is clearer about this in Chapter III of the Analytic of Pure Practical Reason in the *Critique of Practical Reason*.

> [I]f by an incentive we understand a subjective determining ground of a will whose reason does not by its nature necessarily conform to the objective law, it follows... that the incentive of the human will... can never be anything other than the moral law; and... the objective determining ground must at the same time be the exclusive and subjectively sufficient determining ground of action... (Ak: 72; BkR: 74-5).

This difference between the moral law serving as an objective as opposed to a subjective determining ground of an action may be brought out by noticing that an action can be objectively determined by the moral law, but not subjectively. When this occurs, the action conforms to the letter of the law, though is not done as a result of the agent's reverence for the law. The subjective condition would then be "certain impulsions" (Ak: 412; P: 80). In such cases (which Kant refers to as "acting in accordance with duty") sensuous inclinations are the cause of the agents behavior, even though the action is right when measured against the moral law.

> Thus the moral worth of an action does not depend on the result expected from it, and so too does not depend on any principle of action that needs to borrow its motive from the expected result.... Therefore nothing but the representation *of the law* in itself, *which admittedly is present only in a rational being* — so far as it, and not an expected result, is the ground determining the will — can constitute that pre-eminent good which we call moral... (Ak: 401; P: 69).

For Kant then an action which is done "from duty" is one in which "reason infallibly determines the will"; in such actions the will is "in a power to choose *only that* which reason independently of inclination recognizes to be practically necessary, that is, to be good" (Ak: 412; P: 80).

Kant's concept of acting from duty is thus inextricably bound up with the notion of freedom. For to act from duty is to act independently of any inclinations or sensuous impulses, phenomena which are "grounded on empirical laws" (Ak: 427; P: 94). And as Kant tells us in the *Metaphysic of Morals*: "*Freedom* of choice is this independence from sensuous impulse in the determination of choice" (Ak: 213; G: 10). Kant ties the two concepts together explicitly in the second *Critique* saying that

> The essential point in all determination of the will through the moral law is this: as a free will, and thus not only without cooperating with sensuous impulses but even rejecting all of them and checking all inclinations so far as they could be antagonistic to the law, it is determined merely by the law (Ak: 72; BkR: 75).

To be moral is thus to be free, a feat which is accomplished by acting on the basis of that which is independent of the mechanism of nature. It is to act on the "idea of the moral law." Although Kant now and then talks about the idea of the moral law as "the cause genuinely determining our will" (Ak: 407; P: 75), one should be wary of such talk. Kant usually reserves phrases like "determining ground" rather than "cause" to refer to the basis of moral behavior. Kant makes the distinction between "determining ground" and "cause" because this idea of reason is independent of the mechanism of nature and thus need not enter into efficient causal relations. As the determining ground of some actions it does not originate in the natural world, though its effects, human actions, occur in the natural world and have causes in the natural world from which they follow with necessity. According to Kant, then, it is logically possible that some objective events not only have a cause in nature from which they follow in time with necessity, but have a determining ground extraneous to or independent of nature, i.e., the mechanical world.

On the basis of these brief excursions into the theoretical and moral aspects of the Critical Philosophy we can begin filling in the details of Kant's anti-reductionism. Kant holds both that (i) the conceptual materials available to the mechanical point of view are insufficient for the construction of our concept of a living system, thus leaving living phenomena inexplicable in terms of the mechanism of nature, and (ii) the basis of this deficiency lies in the fact that the mechanical conception of nature lacks this idea of a free cause. That Kant holds *both* (i) and (ii) is one of the most closely guarded secrets of the Critical Philosophy.[27] Kant maintains that *all* living processes *involve* the causality of reason, that is, final causes. And final causes are free causes. Thus, to assert that something is a natural purpose entails the view that its internal processes contain events which do not occur with the necessity that is the mark of inorganic occurrences. In living processes there are free causes. To view an expanse of matter as living is to see the matter as manifesting two aspects. One is this reason-like aspect; the other is the effect produced by this reason-like aspect. But the former is operative independently of the causal processes of nature. The movement or activity of this aspect is in some sense free; it is free of the causal effects of any other naturally occurring event.

Two comments are appropriate here. First, Kant's claim that there are free causes in living processes is elliptical. He is actually claiming that living processes must be viewed in terms of the *idea* of a free cause, and all ideas are, as we have seen, non-determining. Second, Kant's "analysis" of the teleological judgment in terms of the idea of free causality is no doubt meant to reflect the formative powers of biological wholes. In living systems, not only are the parts *related mechanically* to the whole, the parts are under the guidance of, or are *formed* by the whole. A formative event, presumably, is a free event. Nevertheless, Kant's view that teleological judgments entail the claim that free causes are present sounds bizarre to the twentieth century ear. In the following section, however, we will watch its strangeness fade by seeing that it is both an attempted characterization of the biology contemporaneous with Kant, as well as the claim that the mechanical conception is unable to fully account for the assertion that internal purposiveness is present in nature. Finally, in Chapter V we will be ready to engage in the final task of this study: Kant's answer to the question "How are teleological judgments possible?"

4. The freedom of vital phenomena

We have seen that Kant's anti-reductionism is grounded in the belief that we cannot fully explain vital processes such as generation and regeneration mechanically, but rather must explain such occurrences in terms of a free cause, the idea of which cannot be constructed with the conceptual resources of the physicochemical point of view. In this final section of the present chapter we will further explore what might be meant by "explaining the vital processes in terms of the idea of a free cause." This will prepare us for the task of Chapter V, where we will consider Kant's argument for the autonomy of biology.

Before we proceed, however, it is worthwhile to indicate one of the ways in which Kant's "analysis" of teleological judgments relates to the architectonic of the Critical Philosophy. It is of interest to note that Kant's concept of a living system is in tune with his pronouncement in the *Critique of Practical Reason* that the notion of freedom "is the keystone of the whole architecture of the system of pure reason" (Ak: 3; BkR: 3).[28] The logical possibility but theoretical impossibility of *freedom* is central to the first *Critique*. In Kant's moral writings moral action is ultimately traced conceptually to *free* action. Moreover, as we have seen, the *free* play of the cognitive faculties is a necessary ingredient of the act of aesthetically judging. For Kant the distinguishing mark of a living system likewise is that its internal processes contain *free* causes. These architectonic considerations are not, of course, meant to be an argument for the correctness of Kant's views. They are rather to be taken as an observation of the elegance of the Kantian system. Kant's philosophy intends to be comprehensive. No one doubts that. But not only has it achieved comprehensiveness (ignoring for a moment its success at getting at the truth), it also has achieved some measure of elegance.

Let us leave such matters and proceed to the final task of this chapter by looking at some remarks made by Kant which are scattered through three papers published between the years 1775-1788.[29] These papers are helpful for our purposes in that they indicate Kant's awareness of a range of biological facts that will in turn shed light on how his anti-reductionist position is to be further understood. The relevant facts are four in number and concern hereditary phenomena.

In the first place, Kant is aware that some characteristics of animals are almost without exception transmitted to their offspring. He realizes that certain characteristics of races of men and sub-species of animals are continually passed on to forthcoming members of the same race or sub-species. Timber wolves, as opposed to the domestic dog which originated from the wolf, always beget offspring with a drooping tail carriage. Members of the Black race of human beings, to use one of Kant's examples, almost always beget children with black skin. Kant is, in the second place, cognizant of the fact that when members of two different races interbreed, the result is a hybrid (*ungleichartige Vermischung*; CH, 95). Black people who breed with White people produce offspring with an intermediate skin color. Moreover, some of Kant's remarks indicate that he comes very close to understanding the difference between dominant and recessive traits.

> Even so I have found that in the marriage between a rational man and a women from a family where insanity is hereditary, but who was herself rational, only one insane child and several clever children result (CH, 94).

Kant thus appears to see that the tendency to produce certain phenotypic traits may be masked by another tendency to produce a different trait. Finally, Kant is aware that there is something like the regulation of gene action involved in the production of some phenotypic traits. Kant notices, for example, that some birds have a moderate layer of feathers in mild climates, but these same birds, should they migrate to a colder climate, develop an additional layer of feathers. He likewise asserts that in a temperate climate wheat has a moderately thick hull, although it produces a very thick hull in a cold climate. Kant realizes then that certain traits of a living thing vary with the environment in which it is placed, a phenomenon which had to wait until the second half of this century to be puzzled out.[30]

These biological facts, sparse as they are, lead Kant to postulate his own hereditary model. He hypothesizes the existence of "*Keime*," which literally translated from German means, "germs," and which it is tempting, because of their functional resemblance, to render as "genes" (DR, 434). These *Keime* are present in the generative force (*Zeugungskraft*; DR, 435) of a living thing. It is the *Keime* which allow for birds to develop a new layer of feathers when they migrate from a temperate to a frigid climate. *Keime*, as part of the generative force, are postulated then as the inner mechanisms for development in future circumstances. In addition to this, they control the permanence of phenotypic traits and are "kept back or unfolded" depending on the situation at hand (DR, 435).

What is surprising, though consistent with the foregoing, is that Kant thinks that *Keime* are not explicable in physicochemical terms.

> So far as I am concerned I trace back all organization of organized beings through heredity and later forms (of these sorts of natural things) according to laws of gradual development from original tendencies.... How this stock [of *Keime*] arose, is an assignment which lies entirely beyond the borders of humanly possible *natural philosophy*, within which I believe I must contain myself (TP, 179).

Indeed, Kant urges that "chance or universal mechanical laws cannot account for adaptations" like the bird's ability to produce an additional layer of feathers in colder climates (DR, 435). Kant accordingly concludes:

> Just as chance or physico-mechanical processes cannot bring forth an organized body, so they cannot develop something like their generative force, that is, something which produces what it communicates to itself, whether it be a special form or a relation of its parts (DR, 435).

Thus, Kant's *Keime*, like the epigenetic capacities of living systems in the Critique of Teleological Judgment, are inexplicable. Just as it is absurd "to hope that another Newton will arise" who will make the production of a mere blade of grass "understandable according to natural laws," explaining the origin of *Keime* is an "assignment which lies entirely beyond the borders of humanly possible *natural philosophy*" ("natural philosophy" meaning "physics").

It is clear that Kant's *Keime*, the Mendelian's genes, and molecular biology's DNA may all be considered as the formative aspect of living things. Insofar as living systems manifest a designer-like level, they are for Kant in some sense "free," and hence irreducible to the physicochemical point of view. But how are we to understand Kant's claim that there are free causes in living processes?

One way of understanding Kant's position is to view it as the claim that we must view living processes in terms of the *idea* of randomness; that is, we might consider these free occurrences as inherently (not just descriptively) random occurrences. In other words, when Kant claims that we must judge living things using concepts of design − concepts which entail the idea of freedom − he means that we must view some biological events as inherently random. Let us briefly look at two contemporary views which rely heavily on the notion of randomness for their construction of the concept of a living system.

We may begin with the views of Ernst Mayr as they appear in his paper "Cause and Effect in Biology."[31] In discussing the programmed information contained in the DNA programs of a zygote Mayr states that

> It is characteristic of these genetic messages that the programming is only in part rigid. Such phenomena as learning, memory, non-genetic structural modification and regeneration show how "open" these programs are.[32]

Moreover, there is a certain amount of randomness in the occurrence of normal biological processes.

> In view of the high number of multiple pathways possible for most biological processes (except the purely physicochemical ones) and in view of the randomness of many of the biological processes... causality in biological systems is not predictive, or at best only statistically predictive.[33]

This causal efficacy of genetic programs, which is only statistical, Mayr calls the "teleonomic" as opposed to the "teleological" aspect of living things, reserving the latter term for cases of intentional human action.

The supposed "openness" of genetic programs, the randomness which Mayr claims to be present in a living system's following alternate routes to achieve a biological goal, is perhaps what Kant is trying to grasp in claiming that there are free causes in biological processes. Perhaps Kant is claiming that the *Keime* of living systems have random effects. Before considering this suggestion, however, let us briefly mention another contemporary version of this construal of a living system.

The same sort of view of living systems is expounded by Robert Ackermann in a paper entitled "Mechanism, Metholodogy and Biological Theory."[34] Ackermann here develops the view that living things are all instances of the concept of an indeterminate mechanism. This position is a denial of reductionism and yet, so he says, is "compatible with sensible physical constraints on the shape of biological theory."[35] Ackermann's view is grounded on the belief

> that deterministic laws are *compatible* with the supposition that the objects characterized by the laws can be analyzed into elements which cannot be characterized in turn by deterministic laws.[36]

The only task left to complete "is to explain how the apparent unpredictability of behavior in certain circumstances can be made compatible with these putative mechanisms."[37] This problem is met by supposing that living things possess mechanisms with random or pseudo-random outputs, and that they can draw on these outputs "when more deterministic mechanisms fail to provide outputs that will permit suitable activity in certain situations."[38] Ackermann gives no explicit examples from biology, but what he has in mind is construing a given metabolic activity of, say, a bacterium, as a result of a randomizing mechanism. By means of this mechanism the procaryotic cells engage in enzymatic production e, f, or g when placed in a certain medium. On this view, a living thing can be likened to an automation that flips a coin or tosses a die when oscillating between alternate activities. Thus, given any instance of behavior we might be able to understand what happened in the organism that led to a certain activity, though we cannot predict this behavior because of its random source.

Both Mayr and Ackermann hold then that many occurrences within living processes are statistical or probabilistic in nature, not just that they are describable in this way. And according to them, herein lies the difference between living and non-living systems. The view itself is, it seems, indefensible. For, ontologically speaking, physics recognizes the inherent randomness of some non-living occurrences. Thus, the presence of statistical phenomena in living processes (if they are present), will not be the distinguishing mark of such processes. But aside from this, could Kant be pushing a view like that of Mayr and Ackermann? Does Kant, in tracing the designer-like quality of living systems to free causes, identify a free cause with a random cause?

It is, of course, clear that a free cause and a random cause need not be identified. For a free cause, particularly as it is described in Kant's moral writings, is one which is not caused by any natural phenomena. Events which occur with only an inherent statistical regularity, however, may nonetheless be effects of causal mechanisms. But it might be thought that Kant wishes to adjust the idea of a free cause to mean something like "a random cause," and use it to demarcate the living from the non-living realm. And such a concept of a free cause would do this for Kant. For he believes that the concept of a physicochemical system is that of a strictly deterministic system. If so, biology would be irreducible to physics in that in biology we "explain" some events as effects of random causes, whereas in physics and chemistry no event is explained in this way.

The difficulty with this interpretation, however, is that there is no textual evidence to support it. It is true that in Kant teleological ascriptions are bound up with "the contingency of nature and its form." But we have seen that to say that something is contingent in nature is for Kant just to say that it is yet to be explained. Moreover, there is just the simple observation that while it is true that biological systems seem to have a teleological aspect, it is far from clear how this could be due to random causes. Yet at this point the same thing might be said of Kant's postulation of free causes.

We can perhaps decipher Kant's meaning by returning to the analysandum of his study: the assertion that internal purposiveness is present in nature. It will be recalled that in Chapter I Kant's concept of internal purposiveness was construed in the following way.

> Where there is a causal relation, x causes y, y possesses internal purposiveness (or y is a natural purpose) if and only if it is *prima facie* legitimate to cite the idea of y to explain the occurrence of x in virtue of (a) x being a part within y, (b) x being the cause of y, and (c) y forming x.

The conditions here together assert that a living thing is a whole comprised of parts which are both the cause of the whole and formed by the whole. In putting it this way Kant is trying to capture certain ways in which we view biological wholes. A tree, he says, produces itself as an individual; it grows. A crayfish regenerates a lost claw. A human being replaces a lost finger nail. These sorts of facts do not rule out a biological whole from being a causal system. Kant says that a living system *is* "the effect of the concurrent moving powers of the parts" (Ak: 77, 407; B: 256). But these facts about living things indicate that we also view biological wholes as forming their parts. And this brings us to Kant's anti-reductionist position. For according to Kant,

> It is entirely contrary to the nature of physico-mechanical causes that the whole should be the cause of the possibility of the parts, rather these must be given in order to conceive from them the possibility of a whole... (Ak: IX, 236; H: 40).

Our concept of a mechanical whole is such that we judge "its form and possibility according to mere mechanical laws, wherein the idea of the effect must be taken not as the ground of the possibility of the cause but rather the reverse..." (Ak: IX, 235; H: 39). But our concept of a living thing is such that we "represent the possibility of the parts (according to their constitution and combination) as dependent on the whole." To Kant the inference is clear. No concept of a whole which forms its parts emerges from the mechanical conception of nature. It follows that biology cannot be reduced to physics, not because biological wholes literally *cause* their parts (for a biological whole is the effect of its parts, and an effect cannot be temporally prior to its cause), but rather because we *explain* the presence and arrangement of their parts in a way which makes sense of the formative powers of a living thing.

As we said previously the sense in which "explain" is used here will be the central topic of the final chapter of this study. Nevertheless, we can finally attach *some* meaning to the Kantian view that there are free causes in living processes. For this claim has a decidely negative import; it is essentially just an affirmation that the mechanical conception of nature and its conception of causality fails to provide a complete characterization of living systems. Kant's doctrine then that "biological causality" involves "free causality" is his way of indicating that teleological judgments are not "possible" in terms of the mechanical point of view. Though these judgments are seen to express the principle of the Second Analogy, a synthetic a priori proposition, this principle fails to capture a significant portion of the content of these judgments. It fails to capture what is meant by living things being "wholes which form their parts." Thus, the claim that there are free causes in living systems has no ontological force. It is rather a transcendental claim, i.e., one concerning the possibility of our judgments.[39] Once we determine how internal purposiveness is possible we will have unpacked the full force of Kant's claim that life involves the "lawless" faculty of freedom.

We might put Kant's position in a way which is no less provincial. His explanatory anti-reductionism is the position that all events are mechanical (or physical), however, biological phenomena are not "explicable" in purely mechanistic terms. Kant's claim that living phenomena contain free causes is but an alternate way of expressing his anti-reductionism. For as Davidson says, "Kant believed that freedom entails anomaly,"[40] i.e., the failure to fall under a causal law. And biological phenomena, to again put it in Davidsonian terms, are "anomalous." Thus, to claim that biological phenomena contain free causes is to understore their anomalogical character. Even though mechanical, living systems must be explained in an anti-mechanistic way. This is a token-token reductionism — an identity of what is biological with what is mechanical — but not a type-type reductionism.

At this point one might object that Kant is simply wrong in maintaining that design- or goal-based concepts cannot be analyzed in a strictly physical way. All of the key features of the concept of a living system, it may be argued, are captured by simply analyzing it, as Kant puts it, as "the effect of concurrent moving powers of the parts." Such an analysis of living systems has been attempted, the objection continues, and with some measure of success.[41] These causal accounts of living systems leave little to be desired as analyses of the concept of a natural purpose. Consequently, any biological event can be sufficiently "explained" by modeling these accounts in whatever sense Kant has in mind. To be more specific, consider a case of regeneration, say, a crayfish (the whole) regenerating (forming) a claw (the part). We can sufficiently describe this process along the lines contained in Mayr's position in a straightforward causal way. When we say that a crayfish regenerates a claw, this is to be understood as one aspect of a system, which we might call the "program" (Kant's *Keime*), controlling the process of regenerating the claw. But that part of the system which is known as the program simply causes (in an admittedly complicated way) the development of the biological part. Thus, a living system is "possible" in terms of the mechanism of nature. When we view living things in such a way that it looks as though the whole is the "cause" of the parts, we are but indicating that there is a program which causes the part to come into existence.

We might even be more specific in our causal characterization, offering what Nagel refers to as a "system-property" view of regeneration.[42] The complete regeneration of the lost claw of the crayfish, the attainment of a normal crayfish is goal G of this process. The mechanisms returning the crayfish to this normal state or homeostasis consist of (to simply matters) two collections of parts. When the crayfish is in its normal state, parts $i_1...i_n$ act as an inhibitor on the construction of or further addition to any claw. If a crayfish loses one of its claws, however, it is replaced *because* of the cessation of the activity of the inhibitor's activities, and the initiation of the activity of another collection of parts, $r_1...r_n$, which engages in the construction of the crayfish's claws. Let K be a variable whose values indicate the rate at which the inhibitor is released and M be the variable whose values indicate the amount of claw construction which occurs. Nagel says that "these variables are independent of (or 'orthogonal' to) each other in the sense that within certain limits the value of either variable at a given moment is compatible with *any* value of the other variable at the *same* moment."[43] If a crayfish loses one of its claws in a fight, the claw is replaced *because* the inhibitor becomes less active and the value of K decreases, while the value of M increases. Once the claw is replaced, the value of K increases and the value of M decreases.

Given this account, one might object to Kant's anti-reductionism by saying that we need not think of a whole forming a part as anything but a complicated causal process. Moreover, though we might explain the activity of regeneration in terms of the formative activities of the whole organism, this is really but a crude way of talking about causal activities within the whole.

How would Kant respond to such an objection? Why is a causal analysis of the concept of a living system incomplete? One response to this objection is that these causal analyses are not flawless. Indeed, the recent literature indicates that serious difficulties are present in each of the various causal analyses.[44] But let us ignore this way of responding and instead look at two others, the first of which is historical in nature and the second philosophical.

One way in which Kant's view might be defended is by pointing to the state of biological thought in which he wrote and realizing that his philosophical characterizations reflect biology in Germany contemporaneous with him.[45] At the time in which the *Critique of Judgment* appeared, there was a general feeling that eighteenth century causal accounts of generation had failed. As a result, many biologists postulated the presence of a vital entity in living processes to account for this process, and saw their role to be that of seeking general descriptive laws of living things. In abandoning these attempts to provide a mechanical account of generation and regeneration, there was in turn a great interest in embryology and physiology, two branches of biology which concentrate on the material structures and their effects in a living system. This dominance of embryology and physiology in biology is reflected by Kant's conception of internal purposiveness. For these branches of biology attend to not only the role played by a structural feature in the biological whole, but what is requisite for this role to be fulfilled by the part. Contemporary biology accordingly saw that a crucial interdependence of their structural components was fundamental to living systems. It emphasized that the proper functioning of any organ was impossible if it was separated from the body to which it belonged. That Kant was aware of this holistic approach to living things is clear.

> It is commonplace that the dissectors of plants and animals, in order to investigate their structure and to find out the reasons why and for what end such parts, such a disposition and combination of parts, and just such an internal form have been given them, assume as indisputably necessary the maxim that nothing in such a creation is *vain*... (Ak: 66, 376; B: 223).

To have characterized a living thing as but the effect of the con-

current moving powers of its parts would not have captured an outlook which emphasized organic wholeness and a biological trait's
dependency on the whole.

This brings us to a second way in which Kant would respond to
these causal analyses of living systems. Though it is true that we
might explain a biological occurrence along the approach laid out by
the system-property view of living things, it does not follow that we
can completely *understand* it in causal terms. And if by "explain" we
mean "produce understanding," then it follows that we cannot completely explain living processes in causal terms. As we shall see in the
following chapter, Kant explicates his second concept of explanation
in terms of the concept of understanding, and in so doing completely
characterizes his explanatory anti-reductionism. For it is his view
that the causal approach is insufficient to understand living processes.

But this raises the question of why, if all processes within nature
are mechanical, they *cannot be entirely understood* mechanically.
That is to say, Kant holds that there are mechanical objects which
cannot be entirely understood in mechanical terms, and if this is not
a contradiction, it is at the very least puzzling. Kant's explanatory
anti-reductionism thus gives rise to a "natural dialectic," whose
solution we must also examine in what follows.

Notes

1. The sense of "explain" here is the sense used in earlier chapters: to explain the occurrence of a phenomenon is to give a causal account of it. Such accounts, moreover, have
a deductive structure. However, Kant also hints that an explanation involves something
else, namely, yielding understanding. I shall have more to say about this in Chapter V.
2. For a very good account of mechanistic biology, as well as its vitalistic opponents contemporaneous with Kant, see Thomas S. Hall's *Ideas of Life and Matter*, Volume I, pp.
218-408 and Volume II, pp. 5-118.
3. See Elizabeth B. Gasking, *Investigations Into Generation*, pp. 70-85 for a good discussion of Maupertuis' views.
4. Cf. Ernest Nagel, *The Structure of Science*, p. 595.
5. I am indebted to Robert Weingard for this point.
6. Kant was familiar with Blumenbach's *Institutiones Physiologicae* (Göttingen, 1786),
translated from original Latin into English by Charles Caldwell (Philadelphia: Dobson,
1795) and *Über den Bildungstrieb* (Göttingen, 1781). Both works were published
before 1790, the year in which the *Critique of Judgment* was published.
   Not only does Kant cite with approval the work of this vitalist, but vitalists cite with
approval the work of Kant. Johannes Müller (1801-58), a staunch vitalist who made
significant contributions to the field of experimental physiology, cites Kant's concep-

tion of a living thing in his *Elements of Physiology*, portions of which are included in *Interrelations: The Biological and Physical Sciences*, Robert T. Blackburn, ed. (Fair Lawn, N.J.: Scott, Foresman and Company, 1966), p. 6.

7. Ibid., pp. 1-2 of the Caldwell translation of this work.

8. Ibid., pp. 34-5.

9. Ibid., p. 588.

10. By "hylozoism" in this context, Kant is alluding to the position held by George L.L. Comte de Buffon (1707-1788) which evoked a *"matier viva"* or living matter, that is, primitive molecules which are both living and indestructable. See his *Histoire des animaux*, Chapter Ten, portions of which are included in Thomas S. Hall's *Ideas of Life and Matter*, pp. 5-17.

11. My argument here requires an important qualification. I am claiming only that the Critical position, to be consistent, must reject vitalism. Indeed, to accept vitalism is to commit what Kant in the first *Critique* calls "The first error which arises from our using the idea of a supreme being in a manner contrary to the nature of an idea, that is, constitutively, and not regulatively only, ...the error of *ignava ratio*" (A 689/B 717). See W.H. Werkmeister, "Kant's Philosophy and Modern Science," *Kant-Studien*, 66 (1975), p. 55 for a compilation of even more evidence that the Kant of the Critical works is not a vitalist. However, passages in the *Opus Postumum* and the lectures on metaphysics in the Pölitz transcription reveal that Kant does flirt with vitalism. Moreover, there is a brief published piece printed as an appendix to Samuel Thomas Sömmerring's *Über das Organ der Seele* (*On the Organ of the Soul*) which seems to favor vitalism. I say "seems to favor vitalism" because we have seen that Kant can consistently praise some features of vitalism without espousing the doctrine (as he does in the *Critique of Judgment* and, apparently, in Section 24 of his *Anthropologie*, Ak: Vol., VII, pp. 161-62). We must in any case distinguish the question of whether *Kant* is consistently anti-vitalistic from the question of whether the *Critical philosophy* is consistently anti-vitalistic. I claim that the latter is so. The former is not as certain.

12. For an informative account of the relationship and influence of Kant's Critical doctrines and Blumenbach's biological theories, see Timothy Lenoir, "Kant, Blumenbach, and Vital Materialism in German Biology," *Isis*, 71 (1980), pp. 77-108. This paper is of particular historical interest in that it advances the thesis that Kant's work was a factor in Blumenbach's drifting away from vitalism in his later career. Also see James L. Larson, "Vital Forces: Regulative Principles or Constitutive Agents?", *Isis*, 70 (1979), pp. 235-49, for an attempt to establish the view "that the use of vital force as a heuristic principle formed a central and coherent strategy within classical vitalism, and that this strategy developed along the lines accurately predicted by Kant" (p. 236).

13. For more information about how these views line up, see Joseph Needham, *A History of Embryology* (Cambridge: Cambridge University Press, 1934), pp. 210-14.

14. It is, of course, possible for one to hold both vitalism and preformation, though there would be little reason to do so.

15. In Chapter V we will consider what Kant means by "nature" here.

16. Cf. D.C. Dennett's description of the physical stance in "Intentional Systems," *Journal of Philosophy*, 68 (1971), p. 88, as well as Hilary Putnam's construal of " 'materialism' in the wider sense" in his "How Not To Talk Aboit Meaning," in *Mind, Language and Reality: Philosophical Papers*, Volume 2 (Cambridge: Cambridge University Press, 1979), p. 120n.

   The present definition, of course, requires the propriety of the use of the concept of causality. But this merely reflects Kant's view that the category of causality is one of those necessary conditions for the "possibility" of knowledge. I will discuss Kant's concept of causality in the following section.

17. In the First Introduction to the *Critique of Judgment* (Ak: V, 215; H: 20) and the

Critique of Teleological Judgment (Ak: 67, 380-1; B: 227-8) may be found additional passages which indicate Kant's ontological reductionist leanings.

18. Cf. John Kemp's discussion in *The Philosophy of Kant* (New York: Oxford University Press, 1968), pp. 114-6.

19. I discuss this incompatibility in more detail in Chapter III, Section 3.

20. Francisco J. Ayala, "Introduction," in *Studies in the Philosophy of Biology*, Francisco Jose Ayala and Theodosius Dobzhansky, eds. (Berkeley: University of California Press, 1974), p. xi. It is to be noted that Ayala is working with the logical empiricist conception of reduction whereby one science is reducible to another (the reducing science) if and only if all of its key terms are *definable* in terms of and all of its laws are *derivable* from the reducing science. It is not far-fetched to view Kant's anti-reductionism in terms of this conception of reduction as well. Since concepts of design are not definable in terms of the physicochemical point of view, and biology makes use of these concepts, the latter is not reducible to the former. However, as we will see in a moment, Kant's anti-reductionism is more accurately construed in terms of a different concept of reduction.

21. Ayala does not hold that the teleological point of view need be construed as the view that living processes involve vital entities or an intelligent designer.

22. Anne M. Faggot, "Causal vs. Teleological Explanation of Behavior" (Ph.D. diss., Stanford University, 1971), p. 10.

23. Kant's explanatory anti-reductionism appears similar, at least in spirit to the anti-reductionist views of Putnam in his two papers "Philosophy and our Mental Life," in *Mind, Language and Reality: Philosophical Papers*, Volume 2 (Cambridge: Cambridge University Press, 1979), pp. 291-303 and "Reductionism and the Nature of Psychology," *Cognition*, 2 (1973), pp. 131-46. Kant's explanatory anti-reductionism is actually much more akin to Donald Davidson's theory of anomalous monism as expounded in his three papers "Mental Events," "Psychology as Philosophy," and "The Material Mind" in his *Essays on Actions and Events* (Oxford: Oxford University Press, 1980).

24. Since living things are entities within the natural world and all natural objects behave in accordance with effective causation, it would appear that we have a contradiction here. But the fact that a biological event is *caused* by some antecedent occurrence does not preclude it being formed by another which is its consequence.

25. It is accordingly ambiguous to say that the principle of the Second Analogy is necessary. For by this, one could mean that (a) every event follows necessarily from its cause, or (b) it is necessary to view nature as conforming to this principle. It is clear that insofar as Kant regards this connection as occurring according to a rule, he believes that the principle is necessary in the first sense. Here the necessity involved is a *real* or *natural* necessity. But Kant also holds that it is necessary to make use of this principle in our attempts to become knowledgeable of nature. This second sort of necessity is a *transcendental* or *epistemic* necessity, of which I will have more to say in the following chapter. Also see Section I of the Appendix of this work.

26. By the "moral law," of course, the "categorical imperative" is meant. The categorical imperative is given several formulations by Kant, but all are intended to be "so many formulations of precisely the same law..." (Ak: 436; P: 103).

27. This is noticed by A.C. Ewing in *Kant's Treatment of Causality* (London: Routledge & Kegan Paul, 1924), pp. 222-35 and R.A.C. Macmillan, *The Crowning Phase of the Critical Philosophy*, pp. 273-4.

28. A recent work on Kant written from this perspective is W. Michael Hoffman's *Kant's Theory of Freedom: A Metaphysical Inquiry* (Washington, D.C.: University Press of America, 1979), although he ignores the aesthetic judgment entirely and says nothing about the freedom of vital phenomena.

29. The three papers are *"Von den verschiedenen Rassen der Menschen"* ("On the Different Races of Men," 1775), *Akademie* edition, Vol. II, pp. 429-43; *"Bestimmung des Begriffs einer Menschenrasse"* ("Determination of the Concept of Human Race," 1785), *Akademie* edition, Vol. VIII, pp. 91-106; and *"Über den Gebrauch teleologischer Principien in der Philosphie"* ("On the Use of Teleological Principles in Philosophy," 1788), *Akademie* edition, Vol. VIII, pp. 159-84. I shall refer to these three works by the abbreviations "DR," "CR," and "TP" respectively and by citing the page in the *Akademie* edition of these works. The translations are mine. A moderately useful work which discusses these papers is Gabrielle Rabel's "Kant as a Teacher of Biology," *Monist*, 41 (1931), pp. 434-70. See also J.D. McFarland, *Kant's Concept of Teleology*, pp. 56-68.

30. This phenomenon is discussed in the classic paper by F. Jacob and J. Monod, "Genetic Regulatory Mechanisms in the Synthesis of Proteins," *Journal of Molecular Biology*, 3 (1961), pp. 318-56.

31. Ernst Mayr, "Cause and Effect in Biology," in *Man and Nature*, Ronald Munson, ed. (New York: Dell Publishing Co., Inc., 1971), pp. 101-16.

32. Ibid., p. 104.

33. Ibid., p. 115.

34. Robert Ackermann, "Mechanism, Methodology, and Biological Theory," *Synthese* 20 (1969), pp. 219-29.

35. Ibid., p. 225.

36. Ibid., p. 226.

37. Ibid.

38. Ibid., pp. 226-7.

39. S. Körner, in *Kant*, p. 211 takes the exact opposite view, saying that Kant's "conjectures, however, that certain phenomena, e.g., the growth of organisms or the affinity between different species, are not susceptible to mechanistic explanation, do not form part of the critical philosophy. They are *obiter dicta* expressing his strong interest in the science of his day and his expectation of its progress." George Schrader also adopts this view in "The Status of the Teleological Judgment in the Critical Philosophy," *Kant-Studien* 45 (1953-54), p. 224.

40. Donald Davidson, *Essays on Actions and Events*, p. 207.

41. A few of these attempts are found in R.B. Braithwaite, *Scientific Explantion*, pp. 319-41; Ernst Mayr, "Cause and Effect in Biology," pp. 101-16; Ernest Nagel, "Teleology Revisited," *Journal of Philosophy* LXXIV (1977), pp. 261-79; Ernest Nagel, *The Structure of Science*, pp. 401-28; Andrew Woodfield, *Teleology* (Cambridge: Cambridge University Press, 1976), pp. 39-106; and Larry Wright, *Teleological Explanations* (Berkeley: University of California Press, 1975), pp. 23-72.

42. Ernest Nagel, "Teleology Revisisted," p. 271.

43. Ibid., p. 273.

44. For a good run-down on the various difficulties which plague causal analyses of the concepts of functionality and goal-directedness, see Peter Achinstein, "Function Statements," pp. 341-67; Ernest Nagel, "Teleology Revisisted," pp. 261-301; and Larry Wright, *Teleological Explanations*.

45. Again see Elizabeth B. Gasking, *Investigations Into Generation*, Chapter 13 and Thomas S. Hall, *Ideas of Life and Matter*, Volume II, pp. 5-118.

CHAPTER V

The autonomy of biology

> For the refractive influence of provincial human interests on the
> construction of teleological explanations is perhaps more often over-
> looked than it is in the case of nonteleological analyses.
>
> Ernest Nagel

1. Kant's projectionism

Kant maintains that a living thing possesses designer-like and designed-like features. Such an object Kant designates a "natural purpose"; alternately expressed, Kant says that internal purposiveness is
present in nature. This is the central kind of teleological judgment.
The goal of Kant's examination of these judgments is the demonstration their "possibility," that is, (a) the discovery of the a priori principle which they express (their metaphysical deduction), and (b) a
defense of this principle (their transcendental deduction). As we saw
in the previous chapter, Kant maintains that the teleological judgment cannot be completely "analyzed" in mechanical terms. In fact,
in offering his "analysis" Kant is to be understood as claiming that
the principle which makes teleological judgments possible – the
teleological maxim – is absent from judgments about mere causal
wholes. It thus becomes clear that his argument that the teleological
maxim is not expressed by our judgments about causal wholes *and*
his defense or transcendental deduction of this principle will complete his proof of the autonomy of biology. Moreover, we shall see
that Kant's argument for the autonomy of biology and defense of
the teleological judgment *are* one and the same; his defense of the
teleological judgment is equivalent to an argument for the autonomy
of biology. It is to these matters that the remainder of this study is
devoted.

There is a way of reading Kant's discussions of living organisms which has dominated interpretations of the Critique of Teleological Judgment.[1] This approach, after recognizing that Kant holds that we lack sufficient evidence to support the claim that living things contain intelligent design in their etiology and that they are in some sense "inexplicable" mechanically, asserts that Kant believes that we, the human observers, project these designer-like and designed-like features onto nature.[2] Kant is often viewed as arguing that since living things have no intelligent designer and cannot be explained mechanically, their teleological aspects are injected into nature by the human subject. One can even construct from Kant's very words in the Critique of Teleological Judgment such an argument.

> (1) [B]ut how purposes that are not our own and that also are not due to nature (which we do not admit as an intelligent being) could or should constitute a special kind of causality, is something which there is absolutely no ground to presume a priori (Ak: 61, 359; B: 205).

> (2) [P]urposiveness, as a principle of the possibility of things of nature, is so far removed from the concept of nature as a series of *necessary* connections that it is more often just that which one calls on to demonstrate the contingency of the same (nature) and its forms. ...[T]hat is, that nature considered as a mere mechanism, could have constructed itself in a thousand different ways without bumping into the unity according to such a principle... (Ak: 61, 360; B: 206).

> (3) [T]here must have preceded then a mental deception which only playfully introduces the concept of a purpose in the nature of things, but does not derive it from objects and experiential knowledge (Ak: 61, 359-60; B: 205).

Since we can neither justify our teleological judgments by citing an intelligent designer nor by mechanistic science, they must be the mere product of, to use Nagel's words, "the refractive influence of provincial human interests."

In Chapter II, we characterized the cognitive activity of the reflective capacity of judgment metaphorically as "soaking" experienced objects with additional concepts, viz., ideas. To say that the idea of design is "projected" onto the world is to offer yet another metaphorical description of the act of judging teleologically. There is, in other words, something to reading Kant's theory of the teleological judgment as arguing (1)-(3). For (3) is, however misleading, a statement of how teleological judgments are possible, as well as an indication of the distinguishing mark of the biological point of view. In this chapter, I hope to show that (3) can be sterilized of its psychological overtones and shown to be the positive side of the claim that living things involve free causes.

To see this, we need to first recall the central claims of the previous chapter. There we saw that one of the features of any objective causal series is that the effect follows from its cause with necessity. And for anything to be objective, it must be such that it is subject to the law of causality. Yet the very concept of a living thing entails that there are free causes within that collection of causal processes regarded as a living organism. As an object in the natural world its processes are all supposed to be subject to the causal principle. But as a biological system it violates this very principle by admitting that there are events which do not follow from some other event with necessity. Thus

> as a concept of a *natural product* it grasps natural necessity and at the same time contingency in the form of the object (in reference to the mere laws of nature) in the same thing as purpose. From this it follows, provided there is no contradiction here, it must contain a ground for the possibility of things in nature and also a ground of the possibility of this nature itself and its reference to something which is not empirically cognizable nature (supersensible)... (Ak: 74, 396; B: 244).[3]

It seems then that Kant's concept of a living system is incoherent.

As we have seen, however, this conflict between necessity and freedom in vital processes is not to be taken as ontological.[4] Kant believes that all natural phenomena behave mechanically. But his claim that living things contain free causes has a different status from the claim that they are mechanical in nature. The language Kant uses in (3) provides a hint at the alternate status of his claim that there are free causes in living processes. Kant tells us that the presence of purposiveness in nature is due to an antecedent mental deception (*Vernünftelei*) whereby the concept of purpose is playfully introduced into the nature of things. What Kant means here is that to judge nature teleologically is to estimate it in terms of the idea of design. And all ideas, as we have seen, are *non-determining*. The playful introduction of the concept of purpose *is* the non-determining use of the idea of design.

The principle that every event necessarily follows from its cause, on the other hand, a "law" to which every object in nature must conform, is a *determining* description of reality. Every causal relation, whether it involves living or non-living processes, is such that the effect follows from the cause with necessity and the cause in turn necessarily follows from some other event. This is part of what we mean by an "objective" feature of the world. Thus, though the concept of a natural purpose entails that a natural purpose is a collection of processes which both contains free causes and events, all of which follow from one another with necessity, these two features do not have the same status. One is non-determining and the other is determining. And in fact this is what Kant tries to show in the Critique of Teleological Judgment.

The general scheme is clear enough, however difficult it looks like it will be to fill in the details. We can begin supplying these details now. Let us first remind ourselves of some of the features common to the idea of design and the category of causality. In viewing the world in terms of both an idea and a category we are basing our judgment upon a priori "representations" (A 320/B 376), representations which are the ground of a principle which is universal and necessary. Moreover, categories, like ideas, are notions which "we ourselves introduce. We would never find them in appearances, had we not ourselves, or the nature of our mind, originally set them there" (A 125). A category, like an idea, to cite a passage from the *Groundwork* quoted earlier, "is likewise no concept of experience, precisely because it carries with it the concept of necessity and so of a priori knowledge" (Ak: 455; P: 123). Both the category and the idea then are projected onto the world. The result is that the features they introduce are importantly *mind dependent*. They are not abstracted from the world, but are the cognitive contributions of the knowing subject. Thus, the non-determining character of the idea of design cannot merely lie in the fact that it is a priori. If this were the case, the idea could not be distinguished from the category.

To understand what is non-determining about the teleological judgment, it is necessary to return once again to Kant's theory of the reflective capacity of judgment. We saw earlier that the reflective capacity of judgment introduces a taxonomic, explanatory, and a non-cognitive systematic unity into nature.[5] Moreover, we saw that this systematic unity is non-determining. The reflective capacity of judgment determines "neither the nature of objects nor the manner of their production..." (Ak: I, 201; B: 8); "such a classification is not ordinary experimental knowledge..." (Ak: V, 215; H: 20). Kant is quite explicit about the non-determining nature of an idea in the construction of a taxonomic systematic unity in the First Introduction of the *Critique of Judgment*. The idea of nature falling into a taxonomic system

> gives no implication of its fitness to a real purposiveness in its products, that is, the production of individual things in the form of systems. For these always could have been, according to intuition, a mere aggregate and yet exist in accordance with empirical laws, which hang together with others in a system of *logical classification*, without having to assume for their particular possibility a real and thereupon appointed concept as a condition of the same, for the ground upon which purposiveness of nature lies (Ak: VI, 217; H: 21).[6]

Kant maintains then that all systematization of concepts, laws, objects, etc. draws lines of separation in the natural world where none may in fact exist.[7] Each system elects certain features as important and essential while neglecting others as unimportant and irrelevant. As time goes by and human standards change, the system of yesterday seems artificial, today's natural. But in fact neither is more "real" than the other. It is this feature of nature's systematicity which is non-determining. Such systematic classifications then are drawn up to fit *our* cognitive needs. However, the practice of systematizing itself is not thereby arbitrary. For the systematizing activity of the human intellect is grounded in an a priori demand that nature be unified. And though Kant only offers the three maxims of reason as vague guidelines by which such systems of nature are constructed, they are nonetheless guidelines. Nevertheless all systems are human contrivances and are neither true nor false.[8]

We now have, however, a way of distinguishing the projection of an idea from the projection of a category. There is no epistemic "choice" involved when a category is projected onto nature: for a claim to be a knowledge claim it *must* conform to the categorial principles. Indeed, part of what objectivity "signifies" is being subject to the principles of the understanding (A 190/B 235). The Critical conception of objectivity − together with other key assumptions − entails the categorial principles.[9] But it *is* possible to have knowledge of nature *and* see it as an aggregate. A knowledge claim does not entail descriptions in which ideas occur. Indeed, the human subject searches for some such description from which we can derive the knowledge claim. Herein lies the difference between the non-determining character of an idea and the determining character of a category. Now the reflective capacity of judgment, as already indicated, also introduces an *explanatory* systematic unity into nature with another product: the teleological judgment. In the following section we shall see that it too is non-determining. The estimation that matter is living, in Kant's words, "has to do altogether with the constitution of our concepts and not with the constitution of things" (Ak: 68, 384; B: 231).

2. Kant's explanatory systematic unity

Insofar as ideas are used in the construction of a systematic unity of nature by the reflective capacity of judgment, we can finish piecing together Kant's "analysis" of teleological judgments. For

> The teleological is no special faculty, rather only the reflective power of judgment in general, inasmuch as it proceeds in theoretical cognition according to concepts at all times, but in consideration of assured objects of nature according to special principles, namely of a mere reflective, not a determinant power of objects. Thus, its employment belongs to the theoretical part of philosophy... (Ak: VIII, 194; B: 31)

The teleological capacity of judgment is part of the reflective power of judgment in general. Thus, we would expect that the teleological capacity of judgment, that faculty of judgment by which we "estimate" that matter is living, also constructs a systematic unity of nature by means of an idea.

This suggestion is in fact borne out by returning once more to Kant's definition of a natural purpose.

> Where there is a causal relation, x causes y, y possesses internal purposiveness (or y is a natural purpose) if and only if it is *prima facie* legitimate to cite the idea of y to explain the occurrence of x in virtue of (a) x being a part within y, (b) x being the cause of y, and (c) y forming x.

In Kant's words, to judge the cause teleologically involves "placing at its basis the idea of the effect of the causality of the cause as the fundamental condition" of this possibility. Particular attention must be paid to the use of "idea" in this context. For in claiming that it is legitimate to cite the idea of the effect, y (the whole), he is saying that the idea of the effect will

> determine the form and combination of all parts not as cause — for here it would be an artificial product — rather as cognitive ground of the systematic unity of the form and combination of all manifoldness which is contained in the given material for him who is estimating it (Ak: 65, 373; B: 220).

Thus, in judging something teleologically we are not stating that an idea *qua* intention has been or is causally operative. It is rather the case that we make a non-determining use of an a priori notion of a certain kind to construct a systematic unity, i.e., to view the cause x as part of a systematic whole. Consequently, it is the judging subject which is "responsible" for the structure and arrangement of the parts. Kant then is punning on the term "idea" when he says that the idea of the whole organism "determines" the order and arrangement of all of its parts. For although his language suggests that by "idea" he means "intention," he in fact is asserting that a particular a priori conception of the judging subject is used to order or systematize empirical particulars. But the idea here is the idea of design.

Kant becomes more specific about this idea which "is to be the ground of the possibility of the natural product" (Ak: 66, 377; B: 223), by expressing it in terms of the teleological maxim of the reflective capacity of judgment: *"An organized product of nature is one in which every part is reciprocally purpose and means"* (Ak: 66, 376; B: 222).[10] Kant tells us that this maxim is, "on account of the universality and necessity which it declares from such a purposiveness, ...an a priori principle, although it be merely regulative..." (Ak: 66, 376; B: 223). Let us see what this maxim implies by teasing out its synthetic a priori character.

There are several difficulties to contend with in arriving at a proper interpretation of the teleological maxim. The first and perhaps not the least significant is that Kant develops his concept of a living system in causal terms. The part is considered to be the cause and the whole organism the effect. But it is a key Kantian doctrine (with which some philosophers are in agreement) that only *events* can enter into causal relations.[11] If so, by "part" Kant should mean "a biological event" and by "whole" that "collection of events which is taken to be a living system." However, he frequently speaks of the part in this context as an organ or structure, and the whole as a living organism, in which case "part" and "whole" seem to refer to an object. Nevertheless, his statement of the maxim forces us to read Kant as meaning an "objective event" by "part" and thus a "collection of objective events" by "whole." For it asserts that one and the same thing is both purpose and means. That is to say, every biological part both *is* a function and *has* a function. But an organ or structure is not something which we can speak of as being a function. We can, however, speak of a biological event as both *being* a function and *having* a function. The muscular activity of the human heart is both an event which *is* a function (say, of a certain occurrence within the central nervous system) and *has* a function (viz., causing the blood to be circulated).

Still, Kant's maxim entails that all parts of a living system are functional and this is entirely unacceptable as it stands. For if Kant means that literally every biological event in a living system both is a function *and* has a function, he is mistaken. The release of pigment contributing to the growth of a harmless brown mole two inches up from my left wrist surely has no function. Kant must mean to imply then that only those parts which are in some sense "natural" to a living system are functional. This, of course, leaves one with the problem of explicating "natural part" in a way that does not make reference to the notion of "any part which is functional," and thus avoiding an "analytic" formulation of this criterion in the Kantian sense of being subject to "the principle of contradiction" (B 12). Kant must accordingly be viewed as asserting that every part of a living system that is also possessed by the *species* of which it is a member is functional.[12] Thus, since our species does not produce a brown mole two inches up from its left wrist, one cannot say of any particular human being that the growth of such a structure is functional.

Yet Kant's formulation of the maxim, even with this qualification, does not allow for the conceptual possibility that a part shared by individual and species has no function. E.g., the release of tissue cells which comprise the human appendix, according to Kant, must be functional. This may bother some, though one's hesitation about accepting it may be lessened somewhat by noticing that even though this maxim entails that the construction of appendix tissue has some function, it does not entail that we *know* what its function is. And after all, it would be rash to say that some part of a living system has no function, simply because the function is presently unknown. For example, a neurophysiologist might contend that the firing of some neuron x in the human brain has no *known* function, not that it has no function.

But there surely are "natural parts," in the Kantian sense, which biologists classify as non-functional (consider for example the development of the wings on an ostrich). Biologists want to recognize the possibility that some biological events have no function. This is easy to sympathize with, and, indeed, Kant perhaps should scale down the teleological maxim in order to capture this aspect of biological methodology. Still Kant's point is a plausible one. The proposition that every "part" of a living thing both is and has a function is neither a mere empirical observation nor true by definition. It is rather a principle of biological methodology. This maxim or principle cannot, of course, disclose the particular purpose of any single part, but rather only asserts that each has a purpose. The teleological maxim is similar in this respect to the principle of the Second Analogy in the first *Critique*. We know that every event in nature follows from some other event in time. But we do not know the particular cause of any given event without recourse to experience.

Consequently, to follow the teleological maxim is to see a given series of events as forming an inter-connected whole by each being a function of another and having a function. In this way we arrive at a concept of systematic unity different from that discussed in the preceding section. For we are no longer concerned with subsuming particulars under kinds. Rather, we view objective events as comprising a complex whole by being related to other parts of the whole in a rather unique way. They are unified in virtue of their being a function and having a function in a certain natural whole. We might call this an "explanatory systematic unity" insofar as it seems to offer a way of *understanding* an objective event. Thus, in claiming that this teleological capacity of judgment is but the reflective capacity of judgment imparting a systematic unity by means of an idea, Kant is to be viewed as asserting that judging teleologically is one way of *understanding* nature. And if, to push this suggestion even further, we recognize a sense of "explanation" where "to explain" is "to bring understanding," then Kant's theory of the teleological capacity of judgment may be read as an attempt at explicating a way of "explaining" living phenomena.

It now becomes apparent that Kant's claim that living processes contain final causes is to be understood as the view that living processes are to be explained in functional terms. Thus, what Kant is groping toward is this. In claiming that biological wholes are the product of the functioning reflective capacity of judgment, Kant is developing a concept of explanation different from the concept of explanation associated with the theory of the *determinant* capacity of judgment. The principle of the Second Analogy guarantees that for *any* naturally occurring event there are causal conditions which lead to its occurrence. The conception of explanation associated with this principle is that of a deduction of the proposition that the event occurs from a description of the specific and general conditions which causally produce the event. In Kant's words, according to this concept of explanation, "To explain demands deducing from a principle which one must therefore clearly know and be able to give an account of" (Ak: 78, 412; B: 261). Every event in nature can in principle be "explained" in this sense.

However, in the Critique of Teleological Judgment and in the second section of the Appendix to the Transcendental Dialectic entitled "The Final Purpose of the Natural Dialectic of Human Reason," Kant is developing a second conception of explanation. In this sense an event is "explained" by allowing its occurrence to be understood. To explain an event in this second sense then is distinct from deducing the proposition that it occurs from a description of its causal conditions. Let us consider the evidence for this reading of Kant's discussions of biological systematic unity and see what is non-determining about the biological point of view.

One of the most striking kinds of claims made by Kant throughout his discussions of organic phenomena is his frequent pronouncement that such phenomena are inexplicable. These kinds of remarks, as already argued, are indicative of the fact that he is working with two concepts of explanation. For when Kant claims that biological phenomena cannot be explained in terms of the mechanism of nature, he does not mean that they cannot be explained in the sense of providing a "deduction." A necessary condition for an event just to be an object of human knowledge is that it be subject to the principle of the Second Analogy. And if this principle is operative, any event is in principle explicable in the sense that the proposition that it occurs is derivable from the description of its causal conditions. For every event in the natural world enters into a causal relation. Thus, Kant's claims that there are inexplicable phenomena in nature must be interpreted as asserting that though we *can* "explain" every natural occurrence in terms of the mechanism of nature in one sense of the word, we *cannot* in another. We cannot, that is, *understand* some aspects of nature by merely providing deductions.[13]

Consider next two key features of the concept of understanding. This concept is such a primitive one that it might appear to forbid any analysis. Nevertheless, when we speak of understanding something, we are at least saying that we "see" how it is connected with something else. For example, suppose that a person is said to understand why in a certain group of gorillas all died. His understanding at least consists in "seeing" that certain facts about the gorillas are connected. He realizes that there is a connection between (a) the particular group of gorillas in question lived in a valley, (b) gorillas characteristically cannot swim; should a gorilla be placed in deep water it will drown, and (c) the valley the gorillas lived in recently became flooded. His realization that certain facts are connected — ignoring for a moment the nature of the *connections* in question — is at least a part of what his understanding consists. *To understand is to "see" that a certain range of facts is united.*

This first feature of what it is to understand something reveals a second. For whether or not a certain fact is understood is contingent upon the person *knowing certain facts*. For example, a child may not understand why a plant grows toward the light. The child knows that (d) the plant grows towards the light, and (e) plants need food, but is ignorant of the fact that (f) plants need light to make food. Whether or not the child understands (d) depends on *his* knowing certain facts and *his* seeing that they are inter-connected. In short, it depends on his cognitive needs at the moment. I am not saying that this is all that there is to understanding, but these are at least two central features of understanding.

Given that there is a sense in which a statement, remark, gesture, etc. is an explanation if and only if it allows the subject to "understand" in this minimal sense, it is possible to view Kant's concept of biological systematicity as a step in the development of a second theory of explanation. For to see an event as biological is to see it systematically connected with other naturally occurring events in a certain way. It is, following the line we are pursuing, to *explain* it in a certain way, i.e., to understand it. In other words, when Kant tells us that in following the teleological maxim a systematic unity is produced, he is telling us that we can understand *and in this sense explain* some events by viewing them as parts which are both purposes and means in a whole. One way of explaining an event in this sense is to show that it is a function and has a function within a collection of events. Thus, following the teleological maxim, insofar as it permits us to see how innumerable causal sequences are interrelated, allows us to understand these events.

If I am right about what Kant is driving at, then just as there are two senses of "explain" in Kant, there are two senses of "understand."[14] For in addition to the present concept of understanding, there is a second more primitive one. It is this latter notion of understanding that Kant is developing with his theory of the determinant capacity of judgment, i.e., the operation of the understanding (*Verstand*). Kant's use of the word "*Verstand*" here is no accident.[15] For the faculty of understanding, as Kant describes it in the first *Critique, systematizes* the raw uncognized data of perception, making our perceptions interconnected, a condition necessary for knowledge. Thus, according to Kant, in this primitive sense of "understand," understanding is necessary for knowledge; but in the second sense, knowledge is necessary for understanding.

In what respect then do we "playfully introduce" biological wholes into nature? What is non-determining about viewing matter as living? The non-determining feature of the biological point of view begins to emerge once we recall what is non-determining about taxonomic systematic unity. To say that nature is taxonomically systematic, i.e., "homogeneous," "specific," and "continuous" and that this is a non-determining description of nature is to say that the human subject draws up systematic classifications of nature according to three guidelines relevant to his needs. Insofar as human needs change, these systematic classifications of objects will change. It is likewise human needs which bring us to view an event teleologically. When we consider the production of a particular event, say the splitting of the hydrogen bonds of a molecule of DNA, we can view this event, following the teleological maxim, as being a function and having a function. But we need not see it in this way. We are free to see it as one link in an indefinite chain of events. Whether or not matter is seen to be living accordingly depends on the judging subject; or to put it in a more contemporary way: it depends on the description we give to a naturally occurring event. It may be that for some of our needs it is best to view this event as being and having a function in a biological whole, i.e., with an eye toward what it contributes to others within this bounded whole. But to view an event teleologically is compatible with its following with necessity from another event in time.

> It may even be that the determinant capacity of judgment might be able to trace everything back to a mechanical explanation, for it can easily be true both that the *explanation* of an appearance, performed by reason under objective principles, may be mechanical, and that for the selfsame object the rule of power of judgment under subjective principles of reflection may be *technical* (Ak: VI, 218; H: 22).

Thus, the teleological maxim will be compatible with the principle of the Second Analogy, though it offers a way of understanding an objective occurrence which is not suggested by the former constitutive principle. The teleological point of view, however, and hence the biological point of view, are non-determining. Teleological judging is but a matter of the human subject linking objective occurrences up in a coherent and systematic fashion due to his needs.

At this point, however, we might ask: "What are these needs which incline us to judge teleologically?" To ask this question is, of course, to raise the question of what justifies our use of the teleological maxim, and is to demand its transcendental deduction. According to Kant, the use of the teleological maxim "is prompted through particular experiences" (Ak: 70, 386; B: 233). He maintains

> Now the concept of a thing as a natural purpose is a concept that subsumes nature under a causality that is only thinkable through reason, in order to judge according to this principle about that which is given of the object in experience (Ak: 74, 396; B: 243).

Kant, in other words, suggests that there are certain natural conditions under which the teleological maxim is applicable and we make use of it when these conditions obtain. We follow the teleological maxim when we

> find among the products of nature special and very widespread genera that contain within themselves such a connection of effective causes within themselves that we must lay at their basis the concept of a purpose, if we also desire simply to employ experience, that is, observation of their inner possibility according to a suitable principal (Ak: IX, 235; H: 39).

But what is this "connection of effective causes" which requires the use of the teleological maxim?

If we skip two centuries ahead we find that a common answer to this question is that function talk arises with respect to those causal systems we can describe in terms of the system-property view of living systems.[16] We talk about "functional" items in causal systems of a certain sort, namely, those that are described by the system-property view.[17] According to the system-property view, a function is any effect of an "item" in a system which brings the system closer to its goal, a concept which is to be analyzed in causal terms.[18] Thus, according to the system-property view, function talk arises with any causal system of this kind; in short, function talk is appropriate to those causal systems which are goal-directed.

According to Kant, on the other hand, functional descriptions are appropriate to parts of those mechanical systems whose parts are considered to be brought about by the whole. We view the parts of such wholes as being related to each other in a way in which the parts of an artifact are interrelated: as being functionally related. And once we recall that these epigenetic wholes are for Kant, ontologically speaking, nothing more than causal systems, we can see a similarity between the ways these two views explain our functional descriptions of some causal systems. According to the system-property view, functional ascriptions are appropriate to parts in causal systems *of a certain sort*. According to Kant, functional ascriptions are appropriate to parts in causal systems which can also be viewed as forming their parts. In fact, it seems that in the end both views assert the same thing insofar as the latter *is* just a causal system of a certain sort. But this raises the question of why the *formative* relation between a living thing and its parts is anything but a certain kind of *causal* relation. This question becomes all the more serious in that Kant readily admits that there is another explanatory maxim of the reflective capacity of judgment, the *causal* maxim. It is to this problem that the final two sections of this essay are devoted.

Before leaving the present section, however, it may be worthwhile to suggest yet another way in which living processes involve the concept of freedom. To see matter as living, to understand a collection of events teleologically, is to view it as being isolated from the rest of the world, as being bounded and independent. From such a perspective, causal processes are seen as having inherent beginnings and inherent termini. Since these wholes are viewed as isolated from the rest of the world (though they are not in fact), they are thought of as *free* of the mechanism of nature which behaves in accordance with the principle of the Second Analogy. Thus, the notion of biological systematic unity allows us to partly decipher what Kant apparently felt was quite important about the power of judgment in general, viz., that

> the capacity of judgment furnishes the mediating concept between the concepts of nature and the concept of freedom. This makes possible the passage from the pure theoretical to the pure practical, from the conformity to law according to the former to the final cause according to the latter, in the concept of a *purposiveness* of nature... (Ak: IX, 196; B: 33).

The teleological capacity of judgment mediates between the understanding and reason. It is like the understanding in that its cognitive contribution is that of unifying a certain manifold. The understanding, of course, unifies and systematizes the manifold of intuition, while the teleological capacity of judgment unifies the manifold of particulars of which we have knowledge. But the teleological power of judgment performs this unifying function through the notion of freedom, the concept of causality of practical reason.

In the previous chapter it was suggested that Kant's conception of a living thing might best be viewed as a development of the organismic standpoint. For when it comes to explaining the production of a biological event, he is both an anti-vitalist and an anti-mechanist. However, it is important to see that his concept of biological systematic unity differs significantly from what is often considered the organismic point of view. For generally speaking, most who defend this perspective hold that though the difference between the organic and the inorganic is not a chemical one, there is nonetheless a difference between the *organization* of organic and inorganic matter.[19] More specifically, living matter is seen as manifesting levels of organization above the physicochemical level, e.g., there is a cellular level and a physiological level to living things, and the behavioral pattern of these higher levels can be viewed as independent of the behavior of the lower level. Kant, on the other hand, agrees that the difference between the two types of matter is not a chemical one, but does not think that it lies in the fact that organic matter has different levels of organization. It rather lies in the fact that the conceptualizing observer conceives of organic matter in a way that is different from the way in which we view inorganic matter. Thus, Kant's organismic position is clearly an "epistemological" claim, whereas the organismic position as it is normally developed, comes dangerously close at making an ontological claim, and nearly collapses into vitalism.

3. A natural dialectic

It is Kant's belief then that biology cannot be *reduced* to physics, but not because biological wholes literally *cause* their parts (for a biological whole is the effect of its parts, and a cause, given the findings of the Second Analogy, must be temporally prior to its effect). Biology is autonomous because we *explain* the presence and arrangement of biological parts in functional terms. This introduces a mode of explanation into biology which is generally reserved for explaining the features of artifacts. Kant's anti-reductionism then is an explanatory anti-reductionism. Biology cannot be reduced to physics because biological wholes, unlike inorganic wholes, are explained teleologically; their parts are explained in functional terms. Insofar as an inorganic whole is just a whole which is caused by its part and does not in turn form its part, it is not susceptible of functional characterization.[20] But so far our analysis of Kant's anti-mechanism in terms of explanatory systematic unity gets one only to the level of what, in the first *Critique*, one would call the "metaphysical deduction" of the teleological judgment, namely, an explanation of what is involved in teleological judgments (or the use of functional explanations) if we make such judgments. At this point the situation is parallel to the first *Critique's* theory of judgment *überhaupt*, and thus Kant's claim to have discovered what pure concepts are, if there are such categories. We must consider then Kant's transcendental deduction of the teleological judgment, i.e., his attempt to guarantee the legitimacy of such judgments insofar as they do instantiate the teleological maxim.

In order to understand Kant's transcendental deduction we need to consider the dialectic of the teleological judgment. The basis of this dialectic lies in the fact that though we can offer an explanation of some events by following the teleological maxim, it is nonetheless possible to explain, i.e., *come to understand* some events by citing their antecedent conditions. There is, in other words, a second explanatory maxim of the reflective capacity of judgment, the causal maxim. The causal maxim, nor surprisingly, asserts: "All production of material things and their forms must be estimated as possible according to mere mechanical laws" (Ak: 70, 387; B: 234).[21] In following the causal maxim we view a naturally occurring event as a link in an indefinite chain of events. We can thus formulate one kind of systematic unity by following the causal maxim and another kind by following the teleological maxim. Yet here the natural dialectic of the teleological capacity of judgment arises. For "Between these necessary maxims of the reflective power of judgment can be found a conflict, consequently an antinomy, which is the basis of a dialectic" (Ak: 69, 386; B: 233). The dialectic arises in this way.

> With this contingent unity of particular laws it can come to pass that the power of judgment proceeds from two maxims in its reflection. The first comes hand in hand *a priori* with the mere understanding. But the other is prompted through particular experiences which bring reason into play, in order to formulate the estimation of corporeal nature and its laws according to a particular principle. Hence it can happen that these two different maxims appear to be unable to stand next to each other, and consequently a dialectic arises... (Ak: 70, 386-7; B: 233).

There thus arises a conflict between the causal and the teleological maxims.[22]

Kant cautions us that the antinomy he is here discussing is not the conflict between the "*Proposition*: All production of material things is possible according to mere mechanical laws" and the "*Counter-*

proposition: All production of material things is not possible accord-
ding to mere mechanical laws" (Ak: 70, 387; B: 234). These two
propositions have the form of principles of the determinant capacity
of judgment, and as such the latter is false of the phenomenal world.
The antinomy facing Kant is rather a conflict between two guidelines
for *understanding* known events in nature.[23] The antinomy arises
when we ask how we are to understand the production, i.e., the
coming-to-be of naturally occurring events, including generation and
regeneration.

But though reason tells us that there are two ways of understanding the production of an event in nature, these maxims of the reflective capacity of judgment "involve no contradiction *in fact*" ("*so
enthält sie in der Tat gar keinen Widerspruch*") (Ak: 70, 387; B: 234;
my emphasis).[24] Though there are two ways to understand an event
in nature, this is irrelevant to the factual questions of whether the
event occurs and what are the causal conditions for its occurrence.
For in following the teleological maxim, we see the event both as
being a function and having a function. And although linguistically
speaking this attribution has an explanatory ring, it is not an explanation of the event in the sense of citing the conditions which lead to
the event's occurrence. Such ascriptions merely indicate a way of
relating an event to others in a whole of which it is part. In Kant's
words

> In fact also nothing is acquired for the theory of nature or the
> mechanical explanation of phenomena of the same through
> effective causes by considering them in relation to each other
> according to the connection of purposes. *The assertion of pur-
> poses of nature in respect to its products*, so far as they consti-
> tute a system according to teleological concepts, *actually
> belongs to the description of nature* which is composed accord-
> ing to a particular guiding thread. *Here reason* indeed performs
> a grand instruction..., but *offers no information on the origin
> and the inner possibility of these forms* which is dealt with by
> theoretical natural science (Ak: 79, 417; B: 266, my emphasis).

Teleological ascriptions are *descriptions* which explain only insofar as
they bring understanding. They do not offer any information on the
origin of the forms we describe purposively. Since they do not make
any ontological claims about the natural world, the conflict between

the two maxims cannot be one between particular facts about the natural world.

Nevertheless, even though following the causal and teleological maxims does not give rise to a factual conflict, it does present a conflict between two different *interests* of reason. There is a conflict here between the two guidelines by which we are to understand nature. For the causal maxim tells us to understand an event by seeing it as the effect of a self-contained whole, an effect which both is a function and has a function. These two guidelines for understanding an event taken together, Kant says, are inconsistent. Is this antinomy resolvable?

Perhaps the first question to ask is why are these two maxims of the reflective capacity of judgment incompatible? But let us resist this question at this point with the promise to deal with it shortly and instead ask *when* do we follow the teleological maxim? The answer is, as we saw in the previous section, that we follow the teleological maxim when we wish to understand certain "products of nature," namely, natural purposes. The crucial question then is *why* do we use functional talk? For only an answer to this question can forestall the objection that such talk, even when possible (or commonplace), is not *necessary*.

Kant's answer to this question actually comes in two stages. At the first stage he claims that if we wish to make sense of the effect forming its cause, we must conceive of it as being *intended*. In order to make sense of a whole which forms its parts we need to think of the whole as possessing a plan or intention that *causes* the parts; we view its parts as being related to each other in a way in which the parts of an artifact are interrelated: as being functionally related. Hence, it is necessary to describe *functionally* (or apply the teleological maxim to) parts of those mechanical systems whose parts are considered to be brought about by the whole.

In this way, the concept of functionality gets introduced into our characterization and indeed is necessary for the complete characterization of living things. But its use does not entail that we regard the etiology of the parts of a living whole as possessing a conscious intention. Rather, biological parts become susceptible of functional characterization; we describe the parts of living things using such expressions as "in order to," "so that," "for the sake of," etc. Function talk gets introduced on the strength of the analogy between the epigenetic capacities of living things and the construction of artifacts.[25] But it does not follow that "function" has the same sense in both cases.[26] When the physiologist uses the term "function," he does not mean "intended effect" (though he might mean an "effect of a whole with designer-like qualities"). He rather views biological parts in ways in which the parts of designed things are often deliberately constructed. Often a part is chosen to be present in a watch because it is believed that the part will *serve as a means to a goal*. Sometimes a part of an artifact, e.g., the secondary pilot jet of a carburetor, is present because its designer believes that *it contributes good to the system*. The *meaning* of functional ascriptions in biology absorbs these facts about intelligent design.[27] Functional ascriptions *seem* to account for the presence of parts in living systems. But they have this explanatory ring only in virtue of the manner in which they are introduced. They do not account for the presence of a part.[28]

Kant now feels that he has justified or ensured the legitimacy of the biologist's use of the teleological maxim. The only way we can make sense of a whole which forms its parts is by conceiving of it as possessing an intention or plan which is the cause of the part. This introduces a mode of explanation into biology which is generally reserved for explaining the features of artifacts. But Kant also thinks that biology cannot be *reduced* to physics; the "physico-mechanical" perspective cannot show how teleological judgments are possible. The mechanistic point of view *cannot*, in short, justify our use of the teleological maxim. Let us now see why.

4. A noumenal question

The situation is this. The teleological maxim has been justified with respect to living things. Two questions remain. First, why is the causal maxim insufficient for understanding living phenomena, and second, what gives biology its autonomy? With these questions answered, the second stage of Kant's deduction will be complete.

Kant's answers to both questions occur in the difficult discussion of the "resolution" of the antimony of the teleological power of judgment. His resolution depends on noticing the particular nature of the human understanding, as initially described in the *Critique of Pure Reason*, though further developed in the *Critique of Judgment*. It is Kant's view that we would find no distinction between "natural mechanism" and "purposive connection" were it not for the fact

that our understanding is *discursive* (Ak: 76, 404; B: 252). According to Kant, a discursive understanding (such as ours) proceeds "from
the universal to the particular" (Ak: 76, 404; B: 252). By this cryp-
tic remark, Kant means that the nature of our cognitive faculties is
such that for knowledge, it is requisite that there be "two entirely
heterogeneous parts," concepts of the understanding and sensory
intuition of objects (Ak: 76, 401; B: 249). In cognition the human
subject must first, to put it metaphorically, grasp the object and
secondly, judge it, deciding what its properties are by subsuming it
under a universal or concept.[29] Our understanding then consists of
"analytic universals," universals which do not indicate the nature of
the particular but rather "must await this determination by the
power of judgment of the subsumption of the empirical intuition (if
the object is a natural product) under the concept" (Ak: 77, 407; B:
255).[30] Our knowledge of an object depends on what sensory data is
given and under what universals it is subsumed. Thus, "what kind
and how very different the particulars may be that can be given" is
purely an accidental matter (Ak: 77, 406; B: 254). In other words,
with an understanding such as ours there is room for the possibility
that an *intuited* particular might possess not only those features it
possesses in virtue of being subsumed under the concepts of our
understanding, e.g., being subject to the law of causality, but additional features for which our understanding has no universal or
concept.

The inability of our understanding to indicate the entire nature of
a particular is contrasted with an understanding that can. The latter
kind of understanding would be "an *intuitive* understanding" (Ak:
77, 406; B: 254), one "which does not go from universal to particular and consequently to the individual (through concepts)" (Ak: 77,
406; B: 254). Such an understanding, which Descartes claims *is*
possessed by humans,

> proceeds from the *synthetic universal* (the intuition of a whole
> as such) to the particular, that is, from the whole to the parts.
> Therefore, this understanding's representation of the whole
> does not contain the contingency of the combination of the
> parts (Ak: 77, 407; B: 255).

In an intuitive understanding there is no distinction between sensibil-
ity and an understanding. Such an understanding would apprehend

and conceptualize a particular simultaneously. In such a cognitive process the intuition of the whole would specify, i.e., entail everything about the particular. An intuitive understanding then would grasp a particular object completely and not be in the position of the discursive understanding, leaving some aspects of objects inscrutable because it lacks a universal under which it can subsume them. Nothing would be contingent, i.e., inexplicable to a being with an intuitive understanding.

Given that an intuitive understanding can grasp a particular in its entirety, it would be able to grasp a particular as both the effect of the concurrent moving powers of its parts and as that upon which the constitution and combination of its parts depend. For an intuitive understand can apprehend all possibilities. And is is logically possible that a cause may come after the effect (though Kant denies that this is possible for an understanding such as ours). Insofar as an intuitive understanding can grasp all logical possibilities, it could grasp how a mechanical whole could "bear itself alternately as cause and effect" without having to fall back on the teleological maxim.

The antinomy between the teleological maxim and the causal maxim then is in principle resolvable in the sense that we can see how it would vanish if things were apprehended by an intuitive understanding. For biological wholes would present nothing problematic to an intuitive understanding. This is, of course, a discussion of what is logically possible; it is, to put it in a Kantian way, a noumenal question, i.e., a question about a logically possible state of affairs as opposed to what we can actually know. As such, Kant's solution to the antinomy may be seen in clearer light. For he brings in the notion of an intuitive understanding to make the point that though it may be "transcendentally" impossible for a whole to be the *cause* of its parts, it is nevertheless logically possible for there to be such a whole. And such a whole could be apprehended by an intuitive understanding. An intuitive understanding could apprehend a whole which forms its parts *without being forced to use the teleological maxim*.

The discursive understanding, on the other hand, can conceivably be unable to know what we humans assert from "common knowledge."[31] As we saw earlier, in discussing his own Critical method Kant maintains that there are two methods by which we may show what is presupposed by our knowledge claims, moral judgments, and teleological assertions: the analytic and the synthetic method. To follow the former is to proceed "from common knowledge to the formulation of its supreme principle," whereas to follow the latter we proceed "from an examination of this principle and its origins to common knowledge in which we find its application." Our common knowledge, then, is either the starting point or consummation of any "Critical" inquiry. And although at the level of common knowledge it is suggested that natural purposes are wholes which literally *cause* their parts, this is, strictly speaking, unknowable by us, though it may be knowable by a being with an intuitive understanding. It is unknowable because we lack a universal or concept under which we could subsume our intuition of biological activity (presuming that there is such intuition). We make up for this deficiency by relying on the teleological maxim.

Now it may be objected that it is simply question begging to deny that we have a concept or universal under which we could subsume our intuition of living processes. The mechanist has already offered us a universal, namely, the concept of causality. Indeed, Kant need only return to his own Second Analogy to find the conceptual materials needed for both understanding living phenomena *and* the reduction of biology.

Nevertheless, Kant's "resolution" of the antinomy of the teleo-logical capacity of judgment might be recast as a response to the mechanist's objection. And perhaps this reformulation is what he has in mind when he contrasts a discursive with an intuitive understanding. For recall that the system-property view claims: (i) biological wholes are causal systems which have features (ultimately susceptible of a causal characterization) different from mechanical wholes, and (ii) function talk is appropriate to these wholes. It is, however, important to notice that even if (i) is correct, it does not *justify* ascribing functions to parts of biological wholes. That is to say, it does not justify viewing living systems in ways in which the parts of designed things are often deliberately constructed: to serve as a means to a goal or to contribute to the good of the system. From the fact that biological systems are just causal systems of a different sort, it is surely unwarranted to infer that function talk is appropriate in the case of biological wholes and *not* inorganic natural wholes.[32] It does not follow that one can begin talking about functions within biologi-cal wholes *qua* causal systems, unless, of course, it is just a definition-al truth that an effect of any part of these systems can be termed a "function." But this does not justify functional descriptions of some natural processes; it just *is* the claim that we regard functions as the effects of parts of such systems. It is, in other words, an "arbitrary choice of a translation scheme."[33] However, the fact that these parts are members of *these* causal systems and not others may be independent of their being functional, if they are functional. The system-property view does not justify ascribing functions to parts of living things.[34]

Kant's "resolution" of the antinomy of the teleological capacity of judgment is, I would suggest, an attempt to make the point that a mere causal account does not tell us why we regard function talk as appropriate to certain causal wholes. It is in this respect that the mechanical point of view does not explain the "possibility" of living things. For our judgment that something is living instantiates the teleological maxim. And the mechanical framework does not explain this "possibility" insofar as it offers no explanation of *why* a mechanical estimation instantiates the teleological maxim; it only asserts that it does. The system-property view does not justify describing some causal wholes teleologically. Kant perhaps comes closest to putting the difference between his and the opposing view in the way we have in the following passage.

> According to the constitution of our understanding, a real whole of nature is considered only as the effect of the concurrent moving powers of the parts. Suppose, then, that we wish not to represent the possibility of the whole as dependent on that of the parts (after the manner of our discursive understanding), but according to the standard of the intuitive (original) understanding to represent the possibility of the parts (according to their constitution and combination) as dependent on that of the whole. In accordance with the above peculiarity of our understanding, it cannot happen that the whole shall contain the ground of the possibility of the connection of the parts (which would be a contradiction in discursive cognition), but only that the *representation* of a whole may contain the ground of the possibility of its form and the connection of the parts belonging to it. Now such a whole would be an effect (*product*) the *representation* of which is regarded as the *cause* of its possibility; but the product of a cause whose determining ground is merely the representation of its effect, is called a purpose. Hence it is merely a consequence of the particular consti-

tution of our understanding that it represents products of
nature as possible, according to a different kind of causality
from that of the natural laws of matter, namely that of purposes and final causes (Ak: 77, 407; B: 256).

It is appropriate to ascribe functions to parts of living systems in
virtue of our viewing living things as wholes which form their parts.
Thus, the system-property view cannot account for function talk in
biology.

Strictly speaking then, Kant does not deny that teleological talk is
translatable into causal talk. What he does deny is that the causal
perspective provides any account of why we would be *interested* in
following the teleological maxim. If, in physiology, "function" implies "an effect which contributes to the whole system," this certainly has a causal force. But the *motivation* behind viewing living things
in this way is not provided by the mechanical framework. And surely
there is a difference between *being justified* in talking about the
world in ways in which parts of designed things are often deliberately
constructed and *showing* that such talk has a causal force. For mechanical wholes in nature are not deliberately constructed. We, therefore, have no right to view them as such. The mechanistic point of
view does not justify our viewing nature in *the way in which we view
designed artifacts*.

It will perhaps be helpful to summarize Kant's rather long and
complex argument. According to Kant, in following the teleological
maxim we view the world in ways in which parts of designed things
are often deliberately constructed, i.e., we see parts as serving as
means to a goal, as contributing to the good of the system, etc. But
one cannot justify viewing undesigned causal wholes in ways in
which parts of designed things are often deliberately constructed, unless one also views these same wholes epigenetically. Since the mechanistic point of view does *not* recognize the presence of epigenetic
wholes in nature, it follows that the mechanistic point of view does
not justify the teleological perspective. Therefore, the teleological
maxim is not "possible" according to the mechanistic point of view.
But biology, by definition, recognizes and studies the existence of
epigenetic wholes. Consequently, biology is autonomous.

Wherein does the necessity of the teleological maxim lie then?
This question is ambiguous. Is Kant claiming that *necessarily* every
part of a living system is both purpose and means, or is he claiming

that *it is necessary* for the human subject *to use* the teleological maxim in estimating living systems? As in the case of the principle of the Second Analogy, he is asserting both (though he does not argue for the former). Every part of a living system is *necessarily* functional (we cannot allow for the conceptual possibility that there may be a biological part that is not functional). But this necessary assertion "must be necessarily presupposed and assumed, for other-wise there would be no thorough-going connection of empirical cognitions in a whole of experience" (Ak: V, 183; B: 20). It is, according to Kant, necessary if we desire to understand how a whole can form its parts. Without it we would only be able to consider a living system as "the effect of the concurrent motive powers of the parts." Notice, however, that Kant is not claiming that we *must* make use of teleological judgments in science. He is only claiming that we must, *if we see nature as containing epigenetic wholes*. But he does not claim that we *have* to see certain wholes as formative. Indeed, Kant would be wrong if he did. For the biochemist is free to view the binding of the amino acyl synthetase molecule to loop I of the tRNA molecule as but an event in an indefinite causal chain of events. In such moments the biochemist is not viewing the event epigenetically. Yet when the biochemist views such an event as a part of an organism's formative activities, the idea of design is a necessary universal for the reflective capacity of judgment in his or her attempt to understand these wholes. And this idea is not re-ducible to the mechanistic point of view.[35] Thus, Kant's defense of teleology is given in terms of our need to unify cognitively a certain kind of whole we *choose* to see in nature and the failure of the causal perspective to justify its teleological talk.

The source of the irreducibility of living things then lies in what Kant calls our "common knowledge." And our common knowledge, oddly enough, is transcendent, for the teleological judgment at the level of common knowledge claims that an effect can be the cause of its cause. The practicing biologist makes this claim as well.

Kant's explanation of the "possibility" of teleological judgments reveals yet another overlooked symmetry in the Critical system. The theistic and vitalistic attempts to explain this possibility fail because there is no intuition, no experiential data which would verify that living things were or are the result of intelligent design. They fail, that is, to supply one of the necessary ingredients of knowledge. The mechanistic hypothesis also fails in its attempt to explain the possibility of teleological judgments because it cannot supply the concept of a whole which is adequate to vital processes. These failures together mirror one of the most famous Kantian doctrines: "Thoughts without content are empty, intuitions without concepts are blind" (A 51/B 75).

Notes

1. This remark is somewhat misleading in that it might suggest that interpretations of the Critique of Teleological Judgment are legion. The fact of the matter is that in comparison with other areas of Kant, this has received very little attention at all.

2. Four works which put the matter this way are Gerd Buchdahl, *Metaphysics and the Philosophy of Science* (Cambridge, Mass.: The MIT Press, 1969), 506-9; R.A.C. Macmillan, *The Crowning Phase of the Critical Philosophy*, p. 299; Andrew Woodfield, *Teleology*, pp. 26-32; and S. Körner, *Kant*, p. 202.

3. This problem is not the "natural dialectic" which arises between two "necessary maxims of the reflective power of judgment" (Ak: 69, 386; B: 233). The Antinomy between the causal maxim and teleological maxim of the reflective capacity of judgment and the apparent contradiction which exists between the principle of the Second Analogy and the claim that there are free causes in living processes are two distinct problems with which Kant is dealing in the Critique of Teleological Judgment. Kant, however, does not make it clear that there are two different dialectical aspects to his answer to the "possibility" of living things. The result is that the Critique of Teleological Judgment is easily viewed as dealing exclusively with one or the other when in fact Kant is here wrestling with both difficulties. I will refer to the antinomy between the two maxims of judgment, along with Kant, as the "natural dialectic of the reflective judgment," and deal with it later in this chapter.

4. To claim that something is or is not an ontological claim in Kant risks a misrepresentation of his views in that he believes that traditional ontology, as a branch of metaphysics, is an empty enterprise (A 247/B 303). However, by an "ontological claim" I mean only a "claim which is subject to verification by science."

5. See Chapter II, Section 2.

6. In this passage and others in the First Introduction and Critique of Judgment we see Kant identifying "purposiveness" with "systematic unity in nature." I do not think that Kant is unsure about how he is using his own terms here. Rather, this identification is the net result of his Critical approach. Kant believes that he is showing that the estimation that purposiveness is present in nature, when "analyzed" Critically, expresses a kind of systematic unity in nature. "Purposiveness" then is the object of Kant's inquiry at the level of "common knowledge"; biological systematic unity is discovered Critically by "pure reason itself" in the form of a "principle." See Chapter II, Section 1.

7. See Ak: VI, 188; B: 25.

8. In *Kant and the Claims of Taste*, p. 50, Paul Guyer claims that the "ascription of the property of systematicity to nature itself is not, in the end, mirrored in the case of aesthetic judgment, for the principle of taste makes no claim about either natural or artificial *objects* of taste, but concerns only ourselves as the makers of such judgments." Guyer, in addition, claims that this taxonomic systematic unity "must be taken to be about nature and not just about our own subjective cognitive needs with

respect to nature" (p. 49). But Guyer argues only that *"at least in form*, the principle of reflective capacity of judgment is about a property which nature itself possesses" (p. 49, my emphasis). In other words, Guyer should have said "only in form."

9. See Section IV of the Appendix of this work.

10. Notice at this point that Kant has also extended his general model of the relationship between a living system and one of its parts to the relationship *between parts* of a biological whole. Kant not only regards the whole as a purpose, but he also regards the *parts* of purposes (as well as means). Not only does a causal relation hold between two parts, but so does a formative relation.

11. This fact was discussed in relation to Kant's concept of relative purposiveness in Chapter I, Section 3 and in Chapter I, Section 4.

12. This reading of the teleological maxim allows us to see how Kant's notion of taxonomic systematic unity ties up with his notion of explanatory systematic unity. For we ascertain which parts of the living things are "natural" to it by following the principle of genera, the principle by which natural kinds are discovered. But one arrives at the relationship between the parts of this whole by following the teleological maxim.

13. Cf. H. Putnam, "Reductionism and the Nature of Psychology," pp. 131-6.

14. The Critical text is supportive here. At A 311/B 367 Kant says "Concepts of reason enable us to conceive [*Begreifen*], concepts of the understanding to understand [*Verstehen*] perceptions." Kant's two concepts of understanding then are the concepts of *begreifen* and *verstehen*.

15. "*Verstand*" in Kant is to be distinguished from the mode of explanation which hermeneutics refers to as "*Verstehen*," though both are related to the verb "*verstehen*" ("to understand"). For a discussion of the origin of *Verstehen*, see Georg Henrik von Wright, *Explanation and Understanding* (Ithaca, N.Y.: Cornell University Press, 1971), Chapter I.

16. This view of living systems was discussed in Chapter IV, Section 4.

17. This claim is made by Nagel in *The Structure of Science*, p. 421 and in "Teleology Revisited," pp. 296-300. Also see Francisco J. Ayala's two works: "Biology as an Autonomous Science," pp. 213-5 and "Teleological Explanations in Evolutionary Biology," *Philosophy of Science*, 37 (1970), pp. 8-9.

18. In "Teleology Revisited," p. 297, Nagel puts it this way: "A functional statement of the form: a function of item i in system S and environment E is F, presupposes (though it may not imply) that S is goal-directed to *some* goal G, to the realization or maintenance of which F contributes."

19. For a more detailed discussion of organismic biology, see Ernest Nagel, *The Structure of Science*, pp. 428-46. For a recent defense of the organismic standpoint so construed, see Ernst Mayr, *The Growth of Biological Thought* (Cambridge: Harvard University Press, 1982), Chapter II.

20. Noice, however, that Kant's anti-reductionism (whether correct or incorrect) has a considerably wider application than he recognized. For given his account of what is meant by a whole forming its parts, entities like social or political groups are not reducible to mere mechanical wholes. To take the case of social groups, one social group frequently leads to or "generates" another; social groups can also replace a lost member; and social groups are largely self-regulating. Whether Kant, upon reflection, would mind this extension of his anti-reductionism is hard to say. But it is clear that given his explication of the concept, a social group is a natural purpose. If so, the teleological maxim applies to these wholes as well.

21. In "The Final Purpose of the Natural Dialectic of Human Reason" of the first *Critique* the causal maxim is referred to as the "second regulative idea of mere speculative reason..., the concept of the world in general" (A 684/B 712). Here the maxim asserts; "we must follow up all conditions of both inner and outer natural appearances, in an

inquiry which is to be regarded as never allowing completion, just *as if* the series of appearances were in itself endless, without any first or supreme members" (A 672/B 700).

22. It is to be noted that Kant, at this point in the Critique of Teleological Judgment, gives an alternate formulation of the *teleological maxim*: "Some products of material nature cannot be estimated as possible according to mere mechanical laws (their estimation requires an entirely different law of causality, namely that of final causes)" (Ak: 70, 387; B: 234).

23. A number of commentators have claimed that Kant resolves this antinomy by making the principle of the Second Analogy regulative. These writers then go on to suggest that Kant's resolution is inconsistent. For example, W.H. Walsh, in "Immanuel Kant," *Encyclopedia of Philosophy*, Paul Edwards, ed. (New York: Macmillan & Co., Inc. and The Free Press, 1967), Vol. IV, p. 320, claims that Kant suggests "that in the end the mechanical and teleological principles stand on the same level, both belonging to the reflective judgment. But it is hard to see how this can be made consistent with the doctrine of the *Critique of Pure Reason*, which ascribes a constitutive force to the concepts of 'pure physics,' or even with the distinction in the *Critique of Judgment* itself between explaining something and merely 'making an estimate' of it." Similarly, H.W. Cassirer remarks: "I find it especially difficult to understand why Kant speaks here of mechanical principles as if they belonged to the faculty of the reflective judgment and not to the understanding" (*A Commentary on Kant's Critique of Judgment*, pp. 347--8). However, given the view of Kant developed here, it is consistent for the causal principle to be both regulative and constitutive provided that we recognize that Kant is working with two concepts of explanation.

24. This way of putting it is ambiguous. Is it a fact that they do not contradict or is there no contradiction between particular facts? Clearly Kant intends the latter. For were he to assert that the teleological maxim and the causal maxim do not in fact contradict, he would be denying his anti-reductionist position.

25. It might be helpful at this point to contrast Kant's concept of an ecosystem, what he calls "external purposiveness in nature," with that of an individual living system. The point of contrast is that the "internal possibility" of the latter depends on our conception of the existence of "a causality according to purposes, a creative understanding" (Ak: 82, 425; B: 275), whereas the "possibility" of external purposiveness is always referred to "the mechanism of nature" (Ak: 82, 426; B: 275). This, as we saw in Chapter I, is due to the fact that here the cause is *not* a part of an epigenetic whole. Thus, the teleological maxim is not necessary for the "possibility" of external purposiveness. Nevertheless, "If we have discovered in nature a faculty which generates products that we can conceive only in accordance with the concept of final cause once, we go further and also may estimate that things which do not make it necessary to seek for their possibility a principle over and above the mechanism of blindly working causes... belong to a system of purposes..." (Ak: 67, 380-1; B: 228-8). We can use the teleological maxim as a heuristic device, seeking out the ecological effects of inorganic events. To take Kant's example once again, we can consider the withdrawal of an entire ocean following the teleological maxim and look as how it affects various levels of organic life.

26. To mistakenly make this inference has led to a common misconception of Kant's views. For example, in "Teleology Revisited," Nagel claims that Kant was an agnostic "about the literal turth of any functional ascription [in biology]....[T]he agnosticism has its source in the assumptions that a process cannot properly be characterized as purposive, if it can be explained on the basis of physicochemical laws, and that an effect of an organic process can count as one of its biological functions only if that process was *intended* or *designed* to produce the stated effect. Agnosticism concerning

the truth of function ascriptions seems to be the price that must be paid for explicating the notion of biological function in terms of conscious intent" (p. 290). But as should be clear from the present discussion, Kant does *not* believe that the notion of biological function may be explicated in terms of conscious intent. Rather he believes that ascribing functional properties to parts of living systems rests on the analogy between epigenetic wholes and the construction of an artifact.

27. See Chapter I, Section 4 for a discussion of the various ways in which the notion of biological function has been explicated.

28. The same point is made in Section I of Robert Cummins, "Functional Analysis," *Journal of Philosophy*, 72 (1975), pp. 741-65.

29. See W.H. Walsh, *Kant's Criticism of Metaphysics*, pp. 12-13 for a helpful statement of Kant's contrast between the discursive and intuitive understanding.

30. As perhaps the reader has noticed, Kant over-uses the term "determine" (*"bestimmen"*). It can, as we have seen, signify a relation between concepts, between objective occurrences, and even (as it is used here) a relation between concept and object.

31. See Chapter II, Section 1.

32. Nagel's claim "A functional statement of the form: a function of item i in system S and environment E is F, *presupposes (though it may not imply)* that S is goal-directed to *some* goal G, to the realization or maintenance of which F contributes" ("Teleology Revisited," p. 297; the former italics are mine), betrays his doubts about this inference as well.

33. Donald Davidson, "Mental Events," p. 223. This, too, is the point of Section II of Robert Cummins' "Functional Analysis."

34. As already indicated, the system-property view is not the only *type* of causal account of living systems; it is, however, the one which has received the most attention in the literature. Other kinds of causal accounts are found in Andrew Woodfield, *Teleology* and Larry Wright, *Teleological Explanation*. Though I shall not argue for the claim here, the system-property view is the least philosophically problematic of such causal accounts.

35. Donald Davison maintains something quite similar about mental events: "When we portray events as perceivings, rememberings, decisions and actions, we necessarily locate them amid physical happenings through the relation of cause and effect; but as long as we do not change the idiom that same mode of portrayal insulates mental events from the strict laws that can in principle be called upon to explain and predict physical phenomena" ("Mental Events," p. 225). Kant's reasons for the irreducibility of "the concepts of biology" to "the concepts of physics" are different from Davidson's reasons for "the irreducibility of psychological concepts" ("Psychology as Philosophy," p. 241). Moreover, there is yet another important respect in which Kant differs from Davidson. Kant's view, to repeat, is that we *must* make use of teleological judgments in science, *if* we see nature as containing epigenetic wholes. He is not stating categorically that we must view the world teleologically. Davidson, however, *does* seem to make such a categorical claim about psychological concepts: "Psychological concepts... cannot be reduced, even nomologically, to others. *But they are essential to our understanding of the rest*. We cannot conceive a language without psychological terms or expressions — there would be no way to translate it into our own language" ("Psychology as Philosophy," pp. 243-4, my emphasis). Davidson's position, in other words, does not allow for the total elimination of the mental, as advocated by Richard Rorty, et al. However, Kant's anti-reductionism *does* allow for the *elimination* of the teleological perspective, provided we cease to view events epigenetically.

Leibniz and the Second Analogy[1]

If the greatness of a piece of philosophical writing is a function of the attention it has received, then Kant's Second Analogy is surely a candidate for the greatest. Kant's Second Analogy has attracted an incredible amount of attention among philosphers. However, to my knowledge, all Anglo-American published discussions of the Second Analogy view it as Kant's response to Hume. It cannot be denied that this approach to the Second Analogy has been of immense value; we have, after all, learned much by reading Kant as stalking Hume. Perhaps this is the best way to read the Second Analogy and the rest of the Transcendental Analytic. Yet we all too often forget that in maintaining his theory of a priori concepts, Kant is also out to set himself apart from Leibniz. The Second Analogy is again no exception, or so I shall argue.[2] More specifically, I will show that there is a Leibnizian side of the Second Analogy: not only does it serve as a response to Leibniz, but the arguments it contains reflect the Leibnizian attempt to establish the principle of sufficient reason.

I

In the first edition the principle of the Second Analogy is "Everything that happens, that is, begins to be, presupposes something upon which it follows according to a rule" (A 189); in the second edition it is "All alterations take place in comformity with the law of connection of cause and effect" (B 232). We are told that something about the principle is necessary. But as the reader of the Second Analogy will testify, when it comes time to pin down the respect in which the principle is necessary, the position begins to blur rapidly.[3]

The way in which the term "necessary" is used in the Second Analogy does not fully come to light until Kant's *Critique of Judgment,*[4] though the Chapter on the Antinomies does begin to tell us something about this use of "necessary." In the third part of Section 9 of the Antinomies Chapter we find a discussion of the concept of "natural necessity" (A 532-58/B 560-86). Here Kant tells us that there is a natural necessity among events according to which each one is "the inevitable outcome of nature" (A 540/B 568). And this "thorough-going connection of all appearances, in a context of nature, is an inexorable law..." (A 537/B 565). What is certain then is that in the Second Analogy Kant is trying to establish the principle of the Second Analogy, along the natural necessity it brings. For our purposes, however, this is enough. We will consider, that is, Kant's reasons for endorsing the principle of the Second Analogy, while maintaining that it entails a "natural necessity" among events.

Anglo-American commentators almost always discuss Kant's Second Analogy in the context of Hume's skeptical remarks about causality. These same students of Kant then usually decide whether Kant has a successful answer to Hume. It is true that Kant's position may serve as a response to Hume, though it is much less than certain that Hume is Kant's primary target. Nonetheless it is useful to locate the Second Analogy in a Humean context in order to see the *way* in which the Second Analogy is an important element in Kant's response to both Hume and Leibniz.

It is Hume's belief that one kind of object of human reason is "matters of fact." A matter of fact is apparently something akin to the logical positivist's state of affairs, e.g., that the bread I consumed for lunch is nutritious, is a matter of fact. And the denial of every matter of fact is possible, since it does not imply a contradiction. Hume tells us that reasoning concerning matters of fact that is not based on the present testimony of the senses or the records of memory is founded on the relation of cause and effect.

> If you were to ask a man, why he believes any matter of fact which is absent: for instance, that his friend is in the country, or in *France*; he would give you a reason; and this reason would be some other fact; as a letter received from him, or the knowedge of his former resolutions or promises. A man, finding a watch or any other machine in a desert island, would conclude that there had once been men in that island. All our reasonings

concerning fact are of the same nature. And here it is constantly supposed, that there is a connexion between the present fact and that which is inferred from it. Were there nothing to bind them together, the inference would be entirely precarious.[5]

The joke is that in the end Hume thinks that the inference is precarious. For the relation of cause and effect is but a relation of constant conjunction. And without *necessary* conjunction, Hume holds that belief in matters of fact cannot be sufficiently justified. For example, consider the belief that *bread is nutritious*. This belief, according to Hume, is founded on the claim that *bread is a cause of my nourishment*. We can put this in the form of an argument.

> (a) Bread brings about or causes bodily nourishment.
> Therefore, (b) Bread is nutritious.

Though this inference looks pretty safe, it is not. For (a) can only mean

> (a′) In the past, nourishment has always followed when bread is consumed.

And (b) surely implies

> (b′) In the future, nourishment will always follow when bread is consumed.

But is clear that

> (a′) In the past, nourishment has always followed when bread is consumed.
> Therefore, (b′) In the future, nourishment will always follow when bread is consumed.

is not a valid argument.

> These two propositions are far from being the same, *I have found that such an object has always been attended with such an effect,* and *I forsee, that other objects, which are, in appearance, similar, will be attended with similar effects.* I shall allow,

if you please, that the one proposition may justly be inferred from the other: I know in fact, that it always is inferred. But if you insist, that the inference is made by a chain of reasoning, I desire you to produce that reasoning (p. 27).

Hume is suggesting then that all reasoning concerning matters of fact is not rational, though not necessarily irrational. What principle do we follow when we adopt a belief concerning a matter of fact?

This principle is CUSTOM or HABIT. For whenever the repetition of any particular act or operation produces a propensity to renew the same act or operation, without being impelled by any reasoning or process of the understanding: we always say, that this propensity is the effect of *Custom* (p. 28).

All reasonings concerning matters of fact then — be they common sense or science — "are a species of natural instincts" (p. 30).

It is clear that were the principle of the Second Analogy "deduced" or justified,

(a) Bread brings about or *causes* bodily nourishment.
Therefore, (b) Bread is nutritious.

would be "rational" or close to it. If the proper analysis of (a) is

(a″) Nourishment is necessarily connected with the consumption of bread.[6]

then

(a″) Nourishment is necessarily connected with the consumption of bread.
Therefore, (b′) In the future, nourishment will always follow when bread is consumed.

looks quite plausible. Kant's Second Analogy *qua* response to Hume then is that the latter inference is to replace the Humean position which moves from (a′) to (b′).[7] Let us now turn to the text of the Second Analogy.

II

Kant's argument in the Second Analogy rests heavily on his claim that "objectivity" depends on the causal principle. More specifically, the gist of the six proofs of the Second Analogy seems to be

> (i) If the principle of the Second Analogy is true, the order in which an event occurs is a necessary order.

> (ii) The only way to distinguish a sequence of mere representations (and they are always sequential) from a sequence of events (*objective representations*) is if the sequence of representations occurs in a necessary order.

> Therefore, (iii) We can distinguish a sequence of events from a sequence of mere representations only if the principle of the Second Analogy is true.[8]

The most explicit statement of (i)-(iii) is found in what I am calling the first proof.

> (i_1) But the concept which carries with it a necessity of synthetic unity can only be a pure concept that lies in the understanding, not in perception; and in this case it is the concept of the *relation of cause and effect*, the former of which determines the latter in time as its consequence....

> (ii_1) I am conscious only that my imagination sets the one state before and the other after, not that the one state precedes the other in the object. In other words, the *objective relation* of appearances that follow upon one another is not to be determined through mere perception. In order that this relation be known as determined, the relation between the two states must be so thought that it is thereby determined as necessary which of them must be placed before, and which of them after, and that they cannot be placed in a the reverse relation.

> (iii_1) Experience itself — in other words, empirical knowledge of appearances — is thus possible only in so far as we subject the succession of appearances, and therefore all alteration to the law of causality....

The second passage is an expression of (ii), and (ii) is the crucial premise of what is a seemingly invalid argument. (ii) addresses the problem of how the manifold, which is always successive, is to be "distinguished" into objective and subjective succession.

> In the synthesis of appearances the manifold is always successive. Now no object is hereby represented, since through this succession, which is common to all apprehensions, nothing is *distinguished* from anything else. But immediately I perceive or assume that in this succession there is a relation to the preceding state, from which the representation follows in conformity with a rule, I represent something as an event, as something that happens; that is to say, I apprehend an object to which I must ascribe a certain determinate position in time – a position which, in view of the preceding state, cannot be otherwise assigned (A 198/B 243).

There is, according to Kant, a need to distinguish [*underscheiden*] objective succession from subjective succession, and the suggestion is that the causal principle can answer this need. Kant puts it this way explicitly in the second proof (A 191/B 236, A 192/B 237, and 193/238), the third proof (A 194/B 239), the fourth proof (A 196/B 241 and A 198/B 244), and the fifth proof (A 199/B 244). In the sixth he says:

> If, then, my perception is to contain knowledge of an event, of something as actually happening, it must be an empirical judgment in which we think the sequence as determined: that is, it presupposes another appearance in time, upon which it follows necessarily, according to a rule. *Were it not so, were I to posit the antecedent and the event were not to follow necessarily thereupon, I should have to regard the succession as a merely subjective play of my fancy; and if I still represented it to myself as something objective, I would have to call it a mere dream* (A 201-2/B 246-7; my emphasis).

How are we to understand Kant's second premise?

Schopenhauer, in *On the Fourfold Root of the Principle of Sufficient Reason*, gives us the following answer.

> [K]ant says that a representation shows objective reality

(which, I suppose, means that it is distinguished from mere phantasms) only through our *recognizing its necessary connexion* — one that is tied to a rule (the causal law) — *with other representations*, and its place in a definite order of the time-relation of our representations. But of how few representations do we *know the place assigned to them by the causal law* in the series of causes and effects! Yet we are always able to distinguish objective representations from subjective, real objects from phantasms.[9]

A few lines later Schopenhauer adds

We might also imagine that, in the above passage, Kant was under the influence of Leibniz, much as he was otherwise opposed to him in the whole of his philosophy, namely when it is observed that exactly similar statements are found in Leibniz's *New Essays Concerning Human Understanding* (Book IV, Chapter 2).[10]

Schopenhauer is asserting that Kant's use of the term "distinguish" signifies that he is putting forward some sort of *phenomenal criterion* for separating subjective successions from objective successions, viz., that succession we see as necessary or *rule-governed*, we know to be objective. Yet surely this interpretation is mistaken. In a passage I have already cited Kant says that "the *objective relation* of appearances that follow one another is not to be determined through mere perception" (B 234). It is wrong then to think that Kant is claiming that the principle of the Second Analogy is true since we need it to use as a criterion for phenomenally sorting out the objective from the subjective.

Still Schopenhauer's interpretation can be modified. We might view Kant to be arguing that the concept of objective experience entails the assertion that we have a certain *skill*, viz., *the ability to discriminate* between objective succession and subjective succession.[11] How we, in fact, make such discriminations, i.e., what phenomenal criterion we really use is a practical problem and of no concern to Kant. But the presence of our ability to do so *proves* the causal principle. However, with this reading, one wonders why Kant thinks that such a discriminating capacity involves, and thereby demonstrates the truth of the principle of the Second Analogy. Surely the presence of what W.H. Walsh calls "ubiquitous causal connections" is not the only way of accounting for such an ability. To be more explicit, many commentators on the Second Analogy have not adequately explained Kant's use of the key term "distinguish," and as a result we are still left with a seemingly invalid argument. We can, however, discover how Kant *is* using the term "distinguish," if we follow up Schopenhauer's reference to Leibniz.

III

Let us consider the discussion in the *New Essays* referred to by Schopenhauer, as well as some remarks Leibniz makes concerning the principle of sufficient reason in his other writings.[12] One finds numerous formulations of the principle of sufficient reason in the Leibnizian corpus.[13] However, in the *New Essays*, Leibniz, through the character Theophilus, formulates it as the principle that "*nothing happens without reason*" (NE: p. 179). In another work not published during his life, Leibniz maintains that empirical truths depend on the principle of sufficient reason (P: p. 112). He also suggests that the principle is provable (L: p. 717), though he on occasion claims that it is axiomatic (P: p. 94 and L: p. 227). Two questions emerge about the passages in question then. In what respect do empirical truths depend on the principle of sufficient reason? How does Leibniz try to justify the principle? Let us consider the former question first.

The passages isolated by Schopenhauer converge on the problem of how one can tell whether a representation is real or imaginary. The character Philalethes, who represents Locke, is of the view "that there is a great difference" between "the idea which is revived by memory" and "that which actually comes to us by the senses" (NE: p. 374). But though Philalethes is confident that there is a great difference between the perceptions one "has 'when he looks on the sun by day, and thinks on it by night'" (NE: p. 374), he does not say what this difference is.[14] Theophilus' position (though not his immediate response) is that

> [W]here objects of the senses are concerned the true criterion is the linking together of phenomena, i.e., the connectedness of what happens at different times and places and in the experience of different men... (NE: p. 374).

This passage suggests that a representation is real if it is linked with other representations: "the true criterion is the linking together of phenomena." And how might phenomena be linked together? In another passage I will cite in a moment it becomes clear that phenomena *linked together in terms of the principle of sufficient reason* is the true criterion of what is real. Thus, Schopenhauer's remarks about Leibniz have initial plausibility. Leibniz does seem to be reacting to the problem of phenomenally distinguishing between the real and imaginary by offering the criterion, in Schopenhauer's words, that we can recognize whether a representation shows objective reality "only through recognizing its necessary connection — one that is tied to a rule (the causal law) — with other representations."

But I think that in order to understand the deeper import of Theophilus' position, we need to focus on his initial response to Philalethes.

> [T]he truth about sensible things *consists* only in the linking together of phenomena, this linking (*for which there must be a reason*) being what distinguishes sensible things from dreams...
> (NE: p. 374; my emphasis).

In this passage it is claimed that the truth about sensible things *consists* only in the linking together of phenomena in terms of *the principle of sufficient reason*. This has the ring of an *account* of truth, not a procedure for distinguishing the real from the imaginary. Moreover, in the same discussion of the *New Essays* Theophilus says

> For it is not impossible, metaphysically speaking, for a dream to be as coherent and prolonged as a man's life. But this would be as contrary to reason as the fiction of a book's resulting by chance from jumbling the printer's type together. Besides, so long as the phenomena are linked together it doesn't matter whether we call them dreams or not, since experience shows that we do not go wrong in the practical steps we take on the basis of phenomena, so long as we take them in accordance with the truths of reason (NE: p. 375).

Leibniz is here telling us that coherent dreams are a possibility; thus, even though a succession of representations appear in a law-like fashion, this does not insure that they are real. Leibniz is admitting that the skeptic cannot be answered on this point except by saying that "we do not go wrong in the practical steps we take" in sorting out the real from the imaginary. In fact, elsewhere in the *New Essays*, Theophilus says that "There is not even a rigorous demonstration to prove that the objects of our senses, and of the simple ideas which the senses present us with, are outside us" (NE: p. 296). If so, it is hardly possible to demonstrate which of our perceptions are real and which are imaginary. Why then would Leibniz be doing that which he confesses cannot be done? Finally, if Leibniz *were* putting forward a criterion for phenomenally distinguishing the real from the imaginary, we would expect him to argue for the principle of sufficient reason in terms of our supposed need to recognize what is real in perception. In fact, he does not, but instead claims that "*the linking together of phenomena* which warrants the truths of facts about sensible things outside us *is verified by truths of reason...*" (NE: pp. 275-5; my emphasis). Leibniz thinks that there is independent justification for the principle of sufficient reason.[15]

Theophilus' remarks then are best construed as an *account* of empirical truth, not a *procedure* for discerning empirical truth. The truth of sensible things consists in their being linked together in terms of the principle of sufficient reason. In claiming that "the true criterion is the linking together of phenomena," the passage upon which Schopenhauer bases his interpretation, Theophilus means something like the "*defining mark,*" not "*phenomenal difference.*" Moreover, there are other Leibnizian doctrines, some of which are expounded in the *New Essays*, which support this reading of the passages in question and which will in turn shed light on the Second Analogy.

Not surprisingly, the doctrines which are relevant are those which concern Leibniz' theory of truth. This, of course, is not the place to consider in detail Leibniz' theory of truth. However, there are some parts of the theory which may be discussed without giving a full-blown exposition of the Leibnizian doctrine, and which will further our understanding of the Second Analogy. According to Leibniz, a proposition is true if and only if the idea, notion, or concept of its predicate is contained in the idea, notion, or concept of its subject (DM: #8, p. 13; NE: p. 486; P: pp. 93-4; P: p. 107).[16] That is to say, the proposition that

The slice of bread I am now eating is nutritious

is true if and only if the concept of being nutritious is contained in the concept of the slice of bread in question. What is important for our purposes is the relationship between Leibniz' theory of truth and the principle of sufficient reason. In the *Theodicy*, or more specifically, in a book review appended to the *Theodicy*, Leibniz says of the principle of contradiction and the principle of sufficient reason: "one may say in a sense that these two principles are contained in the definition of the true and the false" (T: p. 419). That is to say, Leibniz holds that the principle of sufficient reason is a *consequence* of his theory of truth.[17] Leibniz' position, simply put, is that if the proposition

The slice of bread I am now eating is nutritious

is true if and only if the concept of being nutritious is contained in the concept of the slice of bread in question, and the latter concept is by nature complete (another key Leibnizian doctrine[18]), then there is something *in the concept of the subject* which will serve as a reason for the bread's being nutritious. Therefore, with respect to the bread (including its nutritional value), *"nothing happens without reason."* The reason, in short, may be found by analyzing the subject-concept, which is required by the Leibnizian theory of truth.[19]

Given that Leibniz believes that his theory of truth entails the principle of sufficient reason, we can state with some confidence what is going on in the passages which interest Schopenhauer. Leibniz is not, as we suspected, attempting to tell us how we phenomenally distinguish the real from the imaginary. In claiming that

> the truth of sensible things consists only in the linking together phenomena, this linking *(for which there must be a reason)* being what distinguishes sensible things from deams... (NE: p. 374; my emphasis).

he is enunciating his theory of truth and one of its key implications. Given this theory of truth, *the truth of sensible things consists in their conforming to the principle of sufficient reason*. What we have in the *New Essays* is a partial statement of Leibniz' theory of truth; we have a statement of what empirical truth "consists."

In the last paragraph of the Analogies section we find Kant saying:

Had we attempted to prove these analogies dogmatically; had we, that is to say, attempted to show from concepts... that every event presupposes something in the preceding state upon which it follows in conformity with a rule... all our labor would have been wasted. For through mere concepts of these things, analyse them as we may, we can never advance from one object and its existence to the existence of another or to its mode of existence. But there is an alternative method, namely, to investigate the possibility of experience as a knowledge wherein all objects... must finally be capable of being given to us (A 216-7/B 263-4).

For lack of this method, and owing to the erroneous assumption that synthetic propositions, which the empirical employment of the understanding recommends as being its principles, may be proved domgatically, the attempt has, time and again, been made, though always vainly, to obtain a proof of the principle of sufficient reason (A 218-9/B 264).

This should not surprise us. Kant was fully aware of the Leibnizian or dogmatic mode of demonstration and found it unacceptable. The Leibnizian "proof" of the principle of sufficient reason is no exception. So in one respect Schopenhauer was right. Kant was under the influence of Leibniz, if by this it is meant that Kant appears to be responding to Leibniz' attempt to establish the principle of sufficient reason. For according to Kant, the principle of sufficient reason does not follow from Leibniz' theory of truth. All that follows is that with respect to any true proposition, *there must be a reason* for the predicate term being connected with the subject term *in the concept of the subject* of the proposition. Leibniz, in other words, has at best shown that the principle of sufficient reason applies in the conceptual realm only; one cannot go on to draw the ontological conclusion that *nothing happens without reason*. Yet we will see that the "mode of proof" that Kant adopts, one which argues from "the possibility of experience as knowledge wherein all objects... must finally be capable of being given to us" (A 216/B 263), shares important features with the Leibnizian proof. What these features are, will, I hope, become clear in the following section.

156

IV

This reading of Leibniz' remarks about the principle of sufficient reason in the *New Essays* suggests a more Leibnizian interpretation of the Second Analogy. More specifically, it suggests a different way of interpreting Kant's remarks about the need to "distinguish" objective succession from subjective succession. Rather than interpreting these remarks as the claim that the causal principle is true in virtue of the need to phenomenally distinguish objective succession from subjective succession, or that such a discriminating capacity entails the causal principle, Kant should be understood as offering a *definition or account* of objectivity in terms of natural necessity. Such an account, once sufficiently motivated and joined with other key Critical claims, will constitute a "proof" of the principle of the Second Analogy. This approach to the Second Analogy, as we will see, shows that the Critical proof of the Second Analogy shares important features with the Leibnizian proof of sufficient reason.

In the third paragraph of the B edition of the Second Analogy, the paragraph on which commentators agree is the introductory section of the Second Analogy, Kant asserts

> For instance, the apprehension of the manifold in the appearance of a house which stands before me is successive. The question then arises, whether the manifold of the house is also in itself successive.... *Now immediately I unfold the transcendental meaning [Bedeutung] of my concepts of an object.* I realize that the house is not a thing-in-itself, but only an appearance, that is, a representation, the transcendental object of which is unknown. What then am I to understand by the question: how the manifold may be connected in the appearance itself, which yet is nothing in itself (A 190-1/B 235-6; my emphasis)?[20]

This passage makes it clear that in the Second Analogy Kant sees himself to be defining "objective happening" within the constraints of the Critical Philosophy. Kant, that is, tells us in this paragraph that his is an "enquiry" into "what the word 'object' ought to signify with respect to appearances" (A 190/B 235).[21] Later, in the second paragraph of the fourth proof, Kant asks the question: "How, then, does it come about that we... ascribe to them [representations] some mysterious kind of objective reality" (A 197/B 242)? Such remarks indicate that in the Second Analogy Kant is trying to un-pack the concept of objectivity *vis-à-vis* the Critical Philosophy.

The answer is that *objective event* will be defined in terms of *necessary event*. In the Second Analogy Kant is trying to "distin-guish" objective events from subjective events in the sense that he is intent upon indicating what objectivity "signifies." Thus, Kant is arguing that the natural necessity that the principle brings is to be in-cluded in any Critical account of what it is to say that something is an objective event; it is a necessary condition for objectivity. To say that a succession of representations is objective is, for Kant, to say that the order in which they occur is a necessary order. Just as Leib-niz was not attempting to use the principle of sufficient reason to distinguish the imaginary from the real, Kant is not arguing for the principle of the Second Analogy on the basis of our capacity to recognize objective succession; rather, just as Leibniz offers an account of empirical truth in terms of the principle of sufficient reason, so Kant offers an account of objectivity in terms of the prin-ciple of the Second Analogy.

Why, it may be asked, does Kant define objectivity in terms of natural necessity? We find Kant's answer in the passage initially cited in this section. Typically when we "ordinarily" refer to an event as "objective," what we mean is an "event which occurs independent of our experience of it." But notice that Kant cannot make use of this definition of objectivity. For we cannot *know* whether an event which we take to be objective, e.g., a ship moving down stream, actually occurs independently of the subjective conditions of human experience. Kant thinks that he established this in the Transcenden-tal Aesthetic. Consequently, given the phenomena/noumena distinc-tion, the "transcendental meaning" of objectivity must be in terms of natural necessity. However, to establish the principle of the Second Analogy, we must be justified in drawing the subjective/ objective distinction in the first place; there has to be grounds for the distinction between self and world. This Kant hopes he has estab-lished in the Transcendental Deduction of the Pure Concepts of the Understanding. In other words, Kant is offering us the following *valid* argument.

(iv) There is objectivity.
(v) If there is objectivity there is natural necessity.
(vi) The causal principle is true if and only if there is natural necessity.

Therefore, (vii) The causal principle is true.

The argument of the Second Analogy is not (i)-(iii), but (iv)-(vii). To summarize then, we do not find a self-contained transcendental proof of the causal principle in the Second Analogy. We do find a defense of a definition of objectivity in terms of natural necessity. However, if Kant can show that we are justified in maintaining a distinction between self and world, and the net result of the Transcendental Aesthetic — that we know only phenomena, not things as they are in themselves — is correct, Kant will have a formidable argument for the principle of causality, or as he puts it in an explicitly Leibnizian moment, the "principle of sufficient reason" (A 200/B 245).

Kant's position has a markedly Leibnizian flavor. Leibniz gives an account of empirical truth in terms of the principle of sufficient reason and then proceeds to derive the principle of sufficient reason from his general theory of truth. The soundness of the derivation will depend crucially on his general theory of truth. Kant, on the other hand, defines objectivity in terms of the principle of the Second Analogy and then attempts to derive the principle with the help of two central Critical doctrines. Kant's position is now a derivation of the causal principle from the findings of the Transcendental Aesthetic and Analytic of Concepts. But along with Kant's position comes an attack on Leibniz: the principle of sufficient reason is applicable only to the conceptual realm, whereas the principle of the Second Analogy is *at least* applicable to the phenomenal realm.

On this reading, it is still true that the Second Analogy, along with serving as a response to Leibniz, serves as a response to Hume. If Kant is successful, he has shown that we should replace

(a') In the past, nourishment has always followed when bread is consumed.
Therefore, (b') In the future, nourishment will always follow when bread is consumed.

with

(a") Nourishment is necessarily connected with the consumption of bread.
Therefore, (b') In the future, nourishment will always follow when bread is consumed.

as the analysans of

(a) Bread brings about or *causes* bodily nourishment.
Therefore, (b) Bread is nutritious.[22]

But his response, like Leibniz' derivation of the principle of sufficient reason from his theory of truth, is inextricably bound up with
two deep features of the Critical Philosophy: the phenomena/
noumena distinction and his arguments for the possibility of objectivity. This should not disappoint us, for Kant, I am sure, would
have wanted all or none.

Notes

1. The initial draft of this Appendix was written while attending Robert C. Sleigh's 1981 NEH Summer Seminar, "Leibniz Among the Rationalists." I wish to thank Professor Sleigh and the rest of my fellow participants for creating the stimulating environment in which it was developed. A briefer version of this Appendix was read at the 1983 Western Division APA meetings.
2. Even J.N. Findlay, in his *Kant and the Transcendental Object* (Oxford: Clarendon Press, 1981), does not place the Second Analogy in a Leibnizian context (though he does argue that Kant's claim at A 207-9/B 252-5 that "Every cause evinces its causality over a whole period, and produces an effect over that whole period" is a "Kantian principle" which "is Leibnizian," p. 174). This is surprising since Findlay's is a work which has as one of its explicit goals the connecting of the Critical Philosophy "with the thought of Kant's immediate predecessors in the great German scholastic movement which began with Leibniz..." (p. viii).
3. I attempt to clarify and explain Kant's use of the term "necessary" in the Second Analogy in an unpublished paper entitled "The Necessity of the Principle of the Second Analogy."
4. The position in the *Critique of Judgment*, quite briefly, is this. To say that an event's occurrence is necessary is to say that its description is an empirical generalization which is present in a unified system of such generalizations. The empirical generalization in question then may be called a law. The system in which it is present is hierarchical in nature with the lower-order laws being derivable from higher-order laws. "Natural necessity" then is explained in terms of the necessity of causal laws, which is in turn explained in terms of an empirical generalization falling into a "systematic unity" of other generalizations about "the forms of nature." For more details, see my "The Necessity of the Principle of the Second Analogy," Gerd Buchdahl, *Metaphysics and the Philosophy of Science*, p. 501, Paul Guyer, *Kant and the Claims of Taste*, pp. 45-6, and J.D. McFarland, *Kant's Concept of Teleology*, pp. 8-11.
5. David Hume, *An Enquiry Concerning Human Understanding* (Indianapolis: Hackett Publishing Company, 1977), p. 16. All references to Hume are to this work.
6. Kant, of course, would not say that bread and nutrition *per se* are necessarily connected, but rather if bread is in fact a cause of nutrition it is necessarily connected with nutrition. See A 766/B 794.

7. More specifically, this is part of Kant's response to Hume. Kant's response is actually two-fold, viz., (1) the causal principle is a priori, and (2) the causal principle entails a natural necessity among events. I defend this interpretation in "The Necessity of the Principle of the Second Analogy."

8. Those familiar with the literature on the Second Analogy realize that commentators have isolated six separate proofs in the Second Analogy. I individuate the different formulations of the argument in the following way: the second paragraph in the second edition contains the *first* proof, the second to fourth paragraphs of the first edition the *second* proof, the fifth to the seventh the *third* proof, the eighth to the tenth the *fourth* proof, the eleventh to the thirteenth the *fifth* proof, and the fourteenth the *sixth* proof.

9. Arthur Schopenhauer, *On the Fourfold Root of the Principle of Sufficient Reason*, E.F.J. Payne, trans. (La Salle, Illinois: Open Court, 1974), pp. 127-8, my emphasis.

10. Arthur Schopenhauer, *The Fourfold Root*, p. 128.

11. The following interpretations are just several of many in which "distinguish" is left unspecified or means either "have the ability or capacity to discriminate or recognize" or "phenomenally discern."

    W.H. Walsh, in *Kant's Criticism of Metaphysics*, maintains that "in the second Analogy, for example, it is said that we have *the ability to discriminate* objective from subjective successions, and could not exercise this *ability* unless there were ubiquitous causal connections in the experienced world" (p. 102; my emphasis). But later he says that Kant is not "confronting a practical problem," rather "he is in fact concerned with the theoretical foundations of a process..." (p. 140). However, he does not elaborate on this qualification.

    Kemp Smith, in his *A Commentary to Kant's "Critique of Pure Reason"* (New York: Macmillan, 1979), second edition, p. 371, claims that "in the second *Analogy* the problem is how from representations invariably successive a distinction can be drawn between the subjectively determined order of our apprehensions and the objective sequence of events. Or in other words: how under such conditions *we can recognize* an order as given, and so as prescribing the order in which it is apprehended. Or to state the same point in still another manner: how we can distinguish between an arbitrary or reversible order and an imposed or a fixed order..." (the emphasis on "we can recognize" is mine).

    J.D. McFarland, in *Kant's Concept of Teleology* says that "The distinction which can be made between perceptions whose order can be reversed at will, and those whose order is independent of us shows, according to Kant, that we *recognize* the latter succession to be governed by an objective law or rule. In fact, we must *recognize* this if we are to provide our perceptions with the objective reference which they must have if we are to regard them as constituting experience (p. 5; my emphasis).

    T.E. Wilkerson, in *Kant's Critique of Pure Reason* (Oxford: Oxford University Press, 1976), says "*We must find in our experience the clues* that tell us we are perceiving successive objective events. According to Kant the main clue is very simple. When I say that I perceive the sunrise followed by the dawn chorus I am connecting two perceptions in time, for my perception of the sunrise is closely followed by my perception of the dawn chorus. But the vital point, he argues, is that these perceptions are connected in a necessary order..." (pp. 77-8; my emphasis).

    Arthur Melnick, in *Kant's Analogies of Experience* (Chicago: University of Chicago Press, 1973), claims that it is Kant's view that we apply "a rule that *enables us to order* events temporally as asymmetric *on the basis of features* of the events (taking into consideration features of the circumstances)" (p. 90; my emphasis).

12. The following abbreviations are used in referring to the Leibnizian texts:

DM = *Discourse on Metaphysics, Correspondance with Arnauld, and the Monadology*, George Montgomery, trans. (La Salle, Illinois: Open Court, 1962).

NE = *New Essays Concerning Human Understanding*, Peter Remnant and Jonathan Bennett, trans. (Cambridge: Cambridge University Press, 1981).

P = *Leibniz: Philosophical Writings*, G.H.R. Parkinson, ed. (Totawa, N.J.: Rowan & Littlefield, 1975).

L = *Leibniz: Philosophical Papers and Letters*, Leroy E. Loemker, ed. and trans. (Boston: D. Reidel, 1969), second edition.

T = *Theodicy*, E.M. Huggard, trans. (New Haven: Yale University Press, 1952).

Kant was, of course, profoundly aware of Leibniz' *New Essays*. For an excellent account of the extent of this awareness, see Ernst Cassirer, *Kant's Life and Thought*, James Haden, trans. (New Haven: Yale University Press, 1981), pp. 97-100.

13. See DM: 132, DM: 258, L: 698, L: 717, P: 75, P: 88, P: 94, and T: 147-8.

14. Theophilus later helps him out, suggesting that the difference will be in the force and vivacity of the perception, though this far from satisfies Theophilus. See NE: p. 374.

15. See also NE: p. 392 and NE: p. 444.

16. Leibniz apparently meant this theory to be qualified so that it applies to categorical affirmative singular propositions only. See Robert C. Sleigh, Jr., "Leibniz on the Simplicity of Substance," in *Essays on the Philosophy of Leibniz, Rice University Studies* Vol. 63 (1977), No. 4, pp. 109-10. Also see Remnant and Bennett's note on "Analysis," pp. xxiv-v in their translation of the *New Essays*. Lewis White Beck, in *Essays on Kant and Hume* (New Haven: Yale University Press, 1968), p. 85, notes that "Leibniz usually writes that the predicate is included in the subject." And it is true that Leibniz does talk about truth in this way (see P: p. 75, P: p. 87, and P: p. 96). However, it would seem that when Leibniz is speaking crisply, it is the former formulation that he is asserting. For example, at one point in the *New Essays*, where Theophilus is discussing Aristotle's syllogistic logic, he remarks that "instead of saying 'B is C, A is B, so A is C,' Aristotle will express it thus: 'C is in B, B is in A, so C is in A'." Theophilus comments: "This manner of statement deserves respect; for indeed the predicate is in the subject, *or rather the idea of the predicate is included in the idea of the subject*" (NE: p. 486; my emphasis). For more evidence on this point, see Robert C. Sleigh, Jr., "Leibniz on the Simplicity of Substance."

17. Leibniz also states this explicitly at DM: p. 132, L: p. 268, and P: 88. I think that the passage in the *New Essays* on pp. 293-4 is a more or less implicit statement that the principle of reason is a consequence of Leibniz' theory of truth.

18. See P: p. 18 and P: p. 95.

19. We can now understand why Leibniz refers to the principle of sufficient reason as an "axiom." Leibniz tells Arnauld that his (Leibniz') theory of truth "is my fundamental principle, which I think all philosophers ought to agree to, and one of whose corollaries is that commonly accepted axiom: that nothing happens without a reason which can be given why the think turned out so rather than otherwise" (DM: p. 132). Strictly speaking the principle of sufficient reason is a "*corollary*" which is a "*commonly accepted axiom*."

20. Both Beck (*Essays on Kant and Hume*, pp. 145-6) and Smith translate *Bedeutung* in this passage as "meaning". I am willing to go along with this, though we should remember that Kant did not distinguish between meaning and reference, though he does, curiously enough, distinguish between *Sinn* and *Bedeutung* (B 149, A 155/B 194, and A 156/B 195).

162

21. Also see A 225/B 272 and A 376.
22. It may be of interest to note that it is quite usual for French and German Kant scholars to view the Second Analogy in a Leibnizian context and that some Continental Kant scholars believe a Humean interpretation and a Leibnizian interpretation of the Second Analogy will be incompatible. See, for example, Joachin Kopper, "Die Kantliteratur, 1965-1969," *Proceedings of the Third International Kant Congress*, Lewis White Beck, ed. (Dordrecht: D. Reidel, Publishing Company, 1972), pp. 3-15. Kopper maintains that *"Wenn man es überspitzt formulieren wollte, könnte man sagen, es stehen sich in den beiden hier von uns erörterten Richtungen der Kant-Interpreten ein parmenideischer und ein heraklitischer Standpunkt des Verstehens gegenüber"* (p. 13).

Index

164

formative powers, 4, 24-28, 99, 108. *See also* epigenesis
free causes
  in Kant, 97-100
  in vital phenomena, 100-110, 116, 128
function
  artistic, 52-53
  biological, 1-3, 5, 7-8, 17-18, 22-23, 132-133. *See also* teleological maxim

Gasking, Elizabeth B., 78, 110, 113
goal-directedness, 1-3, 5
Guyer, Paul, 55, 56, 140-141, 159

Hall, Thomas S., 78, 110, 111, 113
Harvey, William, 25, 31
Hempel, Carl, 17-18, 31
hereditary phenomena, 101-103
Hoffman, W. Michael, 112
Hume, David, 1, 34, 54, 56, 64, 65, 71-73, 78, 144, 145-147, 159

idea, 12, 20-21, 36-37, 40-46, 52, 55-66, 115
idea of design, 115, 117-119, 120, 123, 125, 139
inexplicability of living things, 5-6. *See also* reductionism, explanatory
intuitive understanding, 134-136, 137-138

Jacob, François, 32, 78, 113
judgment, 30, 57

Kemp, John, 112
Körner, S., 113, 140
Kopper, Joachim, 162

Langer, Susanne K., 56
Larson, James L., 11
Lehman, Hugh, 10, 31
Leibniz, 33, 39-40, 55, 60, 78, 144, 145, 150, 151-155, 158, 159
Lenoir, Timothy, 111
life, 27. *See also* epigenesis, formative powers, and purposiveness, internal
Locke, John, 151

Macmillan, R.A.C., 29, 140
Maupertuis, 81

Mayr, Ernst, 103-105, 108, 113
McFarland, J.D., 29, 159, 160
mechanism, 5, 59-63, 80-83, 85-86, 100, 104, 106-107, 109-110. *See also* reductionism
Meerbote, Ralf, 55
Melnick, Arthur, 160
metaphysical deduction of the teleological judgment, 129-130
method, Critical, 23-24
Monod, Jaques, 113
moral judgments in Kant, 97-99, 112
Müller, Johannes, 110

Nagel, Ernest, 17-18, 31, 78, 108, 110, 113, 114, 115, 141, 142
natural necessity, 145, 157
natural purpose, *see* purposiveness, internal
Needham, Joseph, 111
Newton, 59-60, 80, 81, 103
noumenal questions, 135

objectivity in Kant, 43-44, 53-54, 116, 119, 148-151
organismic standpoint, 87, 128-9

physicalism, *see* reductionism, ontological
physicotheology, 5. *See also* design, argument from
Plantinga, Alvin, 78
possibility of judgments, 37-39, 46, 139
preformation theory, 60-63, 79, 85-86, 87, 111
purposiveness, 7, 9-13, 29-30, 59, 140
  internal, 18-29, 46, 55, 58-59, 105-107, 109, 114, 120
  relative (or external), 13-18, 23, 29, 58-59, 142
  subjective formal, 30, 41
Putnam, 111, 112, 141

random occurrences, 103-105
reductionism, 6, 12, 46-47, 77, 87, 105, 112
  explanatory, 91-92, 94, 99-100, 106-110, 129-140, 141
  methodological, 88-89
  ontological, 87-88, 94-95
reflective capacity of judgment, 40-46, 52, 53, 117-119, 123, 125-126, 139

NIJHOFF INTERNATIONAL PHILOSOPHY SERIES

1. Rotenstreich N: Philosophy, History and Politics – Studies in Contemporary English Philosophy of History. 1976. ISBN 90-247-1743-4.

2. Srzednicki JTJ: Elements of Social and Political Philosophy. 1976. ISBN 90-247-1744-2.

3. Tatarkiewicz W: Analysis of Happiness. 1976. ISBN 90-247-1807-4.

4. Twardowski K: On the Content and Object of Presentations – A Psychological Investigation. Translated and with an Introduction by R. Grossman. 1977. ISBN 90-247-1726-7.

5. Tatarkiewicz W: A History of Six Ideas – An Essay in Aesthetics. 1980. ISBN 90-247-2233-0.

6. Noonan HW: Objects and Identity – An Examination of the Relative Identity Thesis and Its Consequences. 1980. ISBN 90-247-2292-6.

7. Crocker L: Positive Liberty – An Essay in Normative Political Philosophy. 1980. ISBN 90-247-2291-8.

8. Brentano F: The Theory of Categories. Translated by R.M. Chisholm and N. Guterman. 1981. ISBN 90-247-2302-7.

9. Marciszewski W (ed): Dictionary of Logic as Applied in the Study of Language – Concepts / Methods / Theories. 1981. ISBN 90-247-2123-7.

10. Ruzsa I: Modal Logic with Descriptions. 1981. ISBN 90-247-2473-2.

11. Hoffman P: The Anatomy of Idealism – Passivity and Activity in Kant, Hegel and Marx. 1982. ISBN 90-247-2708-1.

12. Gram, MS: Direct Realism – A Study of Perception. 1983. ISBN 90-247-2870-3.

13. Srzednicki, JTJ and Rickey, VF (eds): Leśniewski's Systems – Ontology and Mereology. ISBN 90-247-2879-7.

14. Smith, Joseph Wayne: Reductionism and Cultural Being – A Philosophical Critique of Sociobiological Reductionism and Physicalist Scientific Unificationism. 1984. ISBN 90-247-2884-3.

16. Notturno, MA: Objectivity, Rationality and the Third Realm: Justification and the Grounds of Psychologism – A Study of Frege and Popper. 1984. ISBN 90-247-2956-4.